Anshan Gold Standard Mini Atlas Series

EMBRYOLOGY

System requirement:
- Windows XP or above
- Power DVD player (Software)
- Windows media player 10.0 version or above (Software)

Accompanying Photo CD ROM is playable only in Computer and not in DVD player.

Kindly wait for few seconds for photo CD to autorun. If it does not autorun then please do the following:
- Click on my computer
- Click the **CD/DVD drive** and after opening the drive, kindly double click the file **Jaypee**

Anshan Gold Standard Mini Atlas Series

EMBRYOLOGY

Roopa Kulkarni
Senior Professor and Former Head
Department of Anatomy
MS Ramaiah Medical College
Bengaluru, India

Varsha S Mokhasi
Professor
Department of Anatomy
MS Ramaiah Medical College
Bengaluru, India

Shailaja Shetty
Associate Professor
Department of Anatomy
MS Ramaiah Medical College
Bengaluru, India

Foreword
S Kumar

**Tunbridge Wells
UK**

JAYPEE BROTHERS
MEDICAL PUBLISHERS (P) LTD.
New Delhi

First published in the UK by

Anshan Ltd
in 2010
6 Newlands Road
Tunbridge Wells
Kent TN4 9AT, UK

Tel: +44 (0)1892 557767
Fax: +44 (0)1892 530358
E-mail: info@anshan.co.uk
www.anshan.co.uk

ISBN 13   978-1-905740-22-2

British Library Cataloguing in Publication Data
A catalogue record for this book is available from the British Library

Printed in India by Ajanta Offset & Packagings Ltd., New Delhi

Dedicated to
Our beloved students

**MS RAMAIAH MEDICAL COLLEGE
AND TEACHING HOSPITAL**

MSR NAGAR, MSRIT POST, BENGALURU 560 054 INDIA

ISO 9001:2000 Recognised Institute

Tel: 23605190, 23601742, 23601743, 23605408 Fax: 080-23606213

e-mail: msr_medical@dataone.in Web: www.msrmc.ac.in

Foreword

Knowledge of Embryology—the branch which deals with the development of the human body—is essential for all levels of practitioners of medicine. Clinicians need a comprehensive knowledge of embryology to determine and interpret numerous developmental and congenital defects.

Until recent times such knowledge of well-orchestrated happenings within the womb had to be more by extension of the imagination,, unlike other such subjects as Gross Anatomy and Histology wherein students get to access material for direct study.

Present-day medical education has drawn significant advantages by the application of technology—photomicrographs, models, computer-assisted animations, Internet images.

This handbook prepared from Dr Roopa Kulkarni, Dr Varsha Mokhari and Dr Shailaja Shetty—my faculty colleagues at MS Ramaiah Medical College, Bengaluru, India—is the product of their firm commitment to enable an average undergraduate student to easily understand, analyze and interpret the development.

The communication of exciting ideas in a subject to make it simple and comprehendible is more important than the mere acquisition of detailed facts. "Duty makes us do things well and love makes us do it beautifully".

I am grateful to the Almighty for the privilege of presenting this foreword to this simple presentation, which I am sure will make most medical students and postgraduates confident and comfortable in their approach to this subject.

S Kumar
Principal and Dean
MS Ramaiah Medical College and Teaching Hospital
MSR Nagar, MSRIT Post
Bengaluru, India

Preface

The subject 'Human Embryology' is the basis for understanding human anatomy. It is the most essential subdivision which helps to understand various clinical subjects. Therefore, the content of the subject cannot be reduced. The hurdle in understanding Embryology is to imagine the developmental stages.

In this regard, the authors express their deep sense of gratitude to all those who have inspired and created interest in Human Embryology and the modelers who have taken pains to bring the structures to near reality with proportionate magnification of various parts of the embryo and the fetus.

The development of the human being is so fascinating that the efforts which are made in bringing out this book will not only help in understanding the subject but also kindle the interest in exploring more about the development of the human body.

Roopa Kulkarni
Varsha S Mokhasi
Shailaja Shetty

Acknowledgments

It gives us immense pleasure to express our gratitude to Dr MR Jayaram, Chairman, Gokula Education Foundation for giving us an opportunity to be a part of the organization and carry out this work.

We thank immensely Dr KN Sharma, Executive Director (Academics) and Sri BR Prabhakara, Chief Executive Officer Gokula Education Foundation for their constant encouragement in academic activities of the institution.

We express our deep gratitude to Dr S Kumar, Principal and Dean, MS Ramaiah Medical College, Bengaluru, for his support and concern for the students. His constant encouragement and motivation has made us to bring out this book.

We thank Mr Chetan, who has helped us in modification and digitalization of the photographs.

We are greatly indebted to all the modelers in Anatomy, from 1979 till date for their great contribution to the field of Embryology, without whom the imagination and communication would have become a dream for most of the teachers in Embryology.

We thank all the staff members in the Department of Anatomy for their support and constructive suggestion.

We are extremely thankful to Shri Jitendar P Vij (Chairman and Managing Director), Tarun Duneja (Publishing-Director), H Vasudev, Author Co-ordinator, Bengaluru Branch of M/s Jaypee Brothers Medical Publishers (P) Ltd. for their effort in bringing out our ideas in print form.

We thank one and all.

Contents

1

Human Embryology

INTRODUCTION

Human embryology is that branch of Anatomy which deals with the development of the individual.

The term 'Embryology', appropriately called developmental anatomy, is applied to the various changes which take place during the growth of an animal from the egg to the adult condition.

Embryology may be studied from two aspects:

1. That of ontogeny, which deals only with the development of the individual.
2. That of phylogeny, which concerns itself with the evolutionary history of the animal kingdom.

Terms Used in Embryology

Prenatal Period

Oocyte

Sperm

Zygote

Fertilization age – 1st day

Cleavage

Morula–12 to 32 blastomeres – 3 to 4 days

Blastocyst—5th day

Implantation 6th day

Gastrula—Formation of trilaminar disk

Neurula—Formation of neural tube – 3rd to 4th week.

Embryo—Stage extends up to end of 8th week. Formation of all major structures

Conceptus—Includes embryo and fetal membranes.

Fetus—Unborn offspring

Abortion—Premature stopage of development and expulsion of conceptus

Trimester—A period of 3 calendar months during pregnancy (I, II, III trimesters)

Ist trimester is the critical stage of development.

Postnatal Period

Neonate—First 1 month after birth
Infant—1st year after birth
Childhood—13 months to puberty
Adolescent—11 to 19 years
Adulthood—18 to 21 years
Adults—Upto 40 years
Middle age—41 to 55 years
Old age—Above 55 years.

From fertilization to birth—approx. 280 days are required – equivalent to 40 weeks/10 lunar months.

Age of an embryo or fetus is expressed in terms of weeks or lunar months of Intrauterine Life (IUL).

SPERMATOGENESIS

Spermatogenesis is a term used to describe the formation of spermatozoa from spermatogonia (Fig. 1.1).

It takes place in the seminiferous tubules of the testes.

The wall of the seminiferous tubule has a basement membrane on which the germ cells are situated. These germ cells are supported by columnar cells.

The germ cells are called the **spermatogonia** and the columnar supporting cells are called **Sertoli cells**.

The gametogenic cells at the basement membrane are called **spermatogonia**. They are of two types and they are **type 'A'** and **type 'B'**.

Type 'A' cells is the stem cells which divide into again type 'A' and 'B' cells.

Type 'B' cells divide by **mitosis** into **primary spermatocytes**. These cells are the largest among all the spermatogenic cells and contain darkly stained nucleus.

Each primary spermatocyte divides by **first stage of meiosis** into **secondary spermatocyte**. These cells are smaller. Each secondary spermatocyte contains half the number of chromosomes (22 autosomes and one sex chromosome).

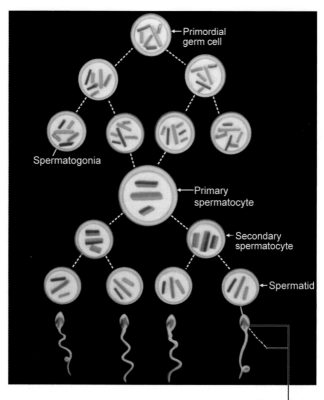

Fig. 1.1: Spermatogenesis

The secondary spermatocytes now divide **by second stage of meiosis** into **spermatids**.

The spermatids undergo **metamorphosis** and become **spermatozoa**.

The conversion of spermatids to spermatozoa is called **spermiogenesis**.

The total duration of spermatogenesis is 64 days. Maturation of the sperms takes place in the epididymis and vas deferens.

The spermatogenesis is controlled by follicle stimulating hormone (FSH) and interstitial cell stimulating hormone (ICSH) of anterior pituitary and adrenal gland.

The mature sperm is about 60 micrometers in length and has head, body and tail.

SPERMIOGENESIS

In this process the spermatids undergo metamorphosis and become spermatozoa.

- The **Golgi apparatus** becomes the **acrosomal cap**.
- The **nucleus** forms the **head of the sperm**.
- **Mitochondria** contribute to body or the middle **piece**.
- **Rest** of the **cytoplasmic contents** form the **body and** the **tail** of the spermatozoon.

Acrosomal cap contains hyaluronidase which dissolves the cumulus oophorus cells and helps the head of the sperm to penetrate through the zona pellucida.

The tail acts as flagellum and propels the sperm forward.

The Sertoli cells are the supporting cells which also nourish, phagocytose the waste products and help in maturation of the sperms. After the sperms mature they become motile.

OOGENESIS

The development and maturation of oocytes in the ovaries is called Oogenesis (Fig. 1.2).

Ovaries are paired organs situated in the pelvis. Each ovary is an almond-shaped structure and consists of an outer cortex and an inner medulla.

The cortex contains the ovarian follicles at different stages of maturation.

At the end of the fetal life, all the oogonia are converted into primary oocytes.

These **primary oocytes** are surrounded by a layer of squamous cells called **follicular cells** and this mass of oocytes and the surrounding follicular cells is called the **primordial ovarian follicle**. Under the influence of follicle

7

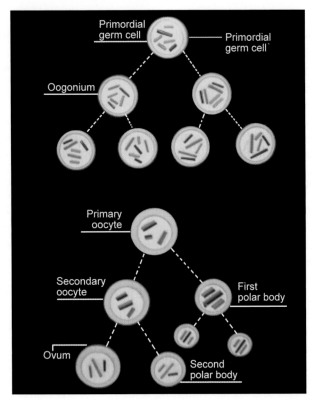

Fig. 1.2: Oogenesis

stimulating hormone the follicular cells gradually become cuboidal and the primordial follicle is converted into **primary follicle**.

The oocyte in the primary follicle divides into **secondary oocyte** and a **first polar body** by **meiosis I**. The follicular cells simultaneously divide by mitosis into many cells which form multiple layers of cells around the secondary oocyte and the 1st polar body. Thus **secondary follicle** is formed. Between the follicular cells and the oocyte there forms a membrane called the **zona pellucida**.

As the follicular cells are multiplying, small spaces appear between the follicular cells which enlarge gradually and coalesce to form a single cavity called the **antrum folliculi** filled with **follicular fluid or liquor folliculi**. Due to the formation of the cavity, the secondary oocyte, 1st polar body along with the zona pellucida are pushed to the periphery and are attached to the inner aspect of the follicular wall. This is called the mature ovarian follicle or the **Graafian follicle** (Fig. 1.3). The secondary oocyte which is in the 1st meiotic division undergoes second meiotic division just before fertilization. If fertilization does not take place, the follicle is extruded along with the menstrual flow before the completion of second meiotic division.

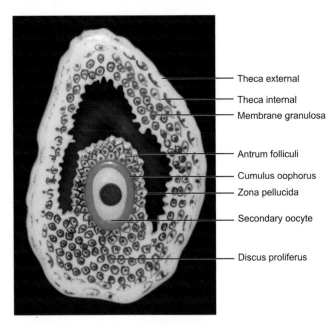

Theca external

Theca internal

Membrane granulosa

Antrum folliculi

Cumulus oophorus

Zona pellucida

Secondary oocyte

Discus proliferus

Fig. 1.3: Graafian follicle

The Graafian follicle consists of secondary oocyte in 1st meiotic division, 1st polar body, both surrounded by zona pellucida. These are attached to the inner wall of the follicle by a group of follicular cells called **discus proligerus**. A layer of follicular cells surround the zona pellucida and are called **cumulus oophorus**. The remaining follicular cells surrounding the antrum folliculi are called the **membrana granulosa**. Around the follicle the stroma condenses and forms **theca**. Theca has an outer fibrous layer (**theca externa**) and an inner cellular layer (**theca interna**). The theca interna cells and membrana granulosa cells secrete estrogen hormone.

Under the influence of luteinizing hormone, the Graafian follicle ruptures following which the secondary oocyte, 1st polar body, zona pellucida, follicular fluid and the cumulus oophorus (now called corona radiata) cells are released into the peritoneal cavity.

The rupture of the follicle and release of ovum is called **ovulation**.

UTERINE CYCLE

Throughout the reproductive life of the female a series of cyclical changes occur in the uterine endometrium, ovary and the mucosa of the vagina. The cycle varies from 25 to 35 days. These cyclical changes are divided into three phases. They are, menstrual, proliferative and secretory phases (Fig. 1.4).

Menstrual Phase

Just before menstruation, three strata can be recognized in the endometrium. They are from deep to superficial—Stratum compactum, Stratum spongiosum and Stratum basale.

Fig. 1.4: Uterine cycle

In stratum compactum there are necks of uterine glands, compactly arranged stromal cells and connective tissue – all lined on the deeper surface, by simple columnar cells.

In stratum spongiosum there are tortuous, dilated uterine glands, stroma in the interglandular space and tortuous capillaries and arterioles. The stromal cells show edematous changes.

In stratum basale are the basal parts of glands, arterioles and stroma. This layer is thin and lies next to myometrium.

Stratum compactum and spongiosum are collectively called stratum functionale. As the corpus luteum (Figs 1.5A and B) regresses in size, stratum functionale undergoes degenerative changes; endometrium shrinks and gets detached from stratum basale. This stage lasts for 3 to 5 days.

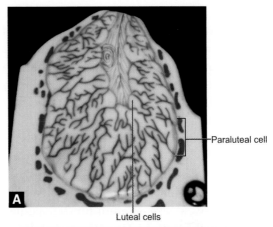

Paraluteal cell

Luteal cells

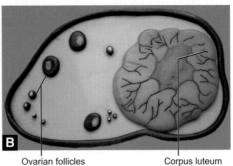

Ovarian follicles

Corpus luteum

Figs 1.5A and B: Corpus luteum

Proliferative Phase

The epithelial lining of the uterine glands in stratum basale grow and replace the detached epithelium. Glands, connective tissue stroma and blood vessels proliferate under the influence of follicle stimulating hormone, progesterone and estrogen. This stage lasts for 10 to 12 days.

Secretory Phase

The uterine glands dilate and becomes secretory. Secretions are rich in nutrients. The stroma proliferate, blood vessels become tortuous and the cells in the stratum spongiosum become edematous the thickness of the enodmetrium becomes three times more than that in proliferative phase, This stage is under the influence of progesterone hormone. This stage lasts for 10 to 12 days.

Corpus Luteum: Plural (corpora lutea) (Figs 1.5A and B)

The term is derived from Latin word which means 'yellow body'.

It is an endocrine part of the ovary which has got a short life.

The function of it is to produce a hormone called 'progesterone' that is required to maintain the thickness of the endometrium at the time of implantation and growth of the embryo.

It is formed when the Graafian follicle ruptures and releases the secondary oocyte (ovulation) on the 14th day of 28 days' menstrual cycle. The rupture of mature ovarian follicle (graafian follicle) and formation of corpus luteum are under the influence of luteinizing hormone.

Following the release of the secondary oocyte along with the cumulus oophorus, zona pellucida and follicular fluid, the follicle forms 'corpus hemorrhagicum' due to rupture of adjacent capillaries and accumulation of blood into the lumen of the ruptured follicle. Then the blood coagulates which is gradually removed by the phagocytosis and sprouting of the ruptured blood capillaries. These newly growing blood vessels nourish the cells of membrana granulosa. During this process, the cells enlarge, accumulate yellowish pigment inside and lead to increase in the size. Due to its yellowish appearance the whole cell mass is called 'Corpus luteum'. It is usually large in size and the diameter varies from 2 to 5 mm in diameter. The yellowish cells are called 'luteal cells' and they secrete the hormone 'progesterone'.

The cells of theca interna of mature ovarian follicle are now called paraluteal cells and they continue to secrete estrogen.

Functions

The corpus luteum is essential for establishing and maintaining pregnancy in females by maintaining the growth and thickness of endometrium.

Fate

When the ovum is not fertilized: If the ovum is not fertilized, the corpus luteum stops secreting progesterone and decays (after approximately 14 days in humans). It is called corpus luteum of menstruation. It then degenerates into corpus albicans as it appears whitish in color. It is a mass of fibrous tissue.

When ovum is fertilized: If the ovum is fertilized and implantation takes place, the trophoblastic cells secrete HCG (human chorionic gonadotropin) which stimulates the corpus luteum to secrete more of progesterone, thus helps in maintaining the thickness of endometrium. It is therefore called 'corpus luteum of pregnancy'. It usually functions up to third-forth month and gradually regresses when placenta takes over the function of secreting progesterone. Sometimes it remains up to the end of gestation period (9th month of pregnancy) and then regresses to for 'corpus albicans'.

Prostaglandins cause degeneration of corpus luteum of pregnancy and cause abortion of the fetus.

CAPACITATION OR ACTIVATION OF SPERM

Capacitation or activation of sperm is a calcium-dependent event which involves extensive changes in the sperm.

1. The surface membrane covering the head of the sperm fuses at many points with the underlying acrosomal membrane creating a vesiculated appearance and exposing the enzymatic contents of the acrosomal

vesicle and also the inner acrosomal membrane to the exterior.

2. The tail beat changes from regular wave-like flagellar beats to "whip lashing beats" that push the sperm forward in vigorous lurches.

3. The surface membrane of the middle and posterior half of the sperm head fuses to the surface membrane of the ovum.

Destabilization of the surface membranes may be critical. To summarize:

- **Hypermotility of sperms**.
- Proteolytic enzymes in female reproductive tract **strip off the glycolytic coat** on the acrosomal cap of the sperm.
- This decreases membrane stability.
- Increases membrane permeability.
- Increases internal calcium levels.
- Increases cAMP.
- Increases phosphorylation of proteins.
- Increases tyrosine kinase.

In the female reproductive tract, sperms undergo capacitation.

FERTILIZATION

Following ovulation, the ovum with its cumulus oophorus cells are picked up by the fimbriae of the fallopian tube. The ovum has formed the first polar body. It remains in the ampulla portion of the tube and is viable for about 18 to 24 hours.

If fertilization does not occur, the ovum disintegrates and is destroyed by the tube.

Sperm will remain viable in the female reproductive tract for about 48 hours, although this can be quite variable.

Sperms present in the ampulla meet the cumulus oophorus mass and penetrate by chemical and mechanical means to reach the zona pellucida.

Once sperm penetrates the zona pellucida, the second polar body is formed, and the nuclear material of the sperm enters the vitelline membrane.

The diploid chromosome number is re-established, and mitotic cell division can now occur.

Sperms stay behind in the oviduct for at least 8-24 hours.

Purposes of Fertilization

1. Completion of second stage of meiosis in oocyte.
2. Restoration of chromosomal compliment of the species.

3. Gives rise to a new individual of its kind.
4. Determination of the sex of the individual.
5. At fertilization a calcium wave is initiated which also prevents entry of other sperms.
6. Initiation of cleavage (rapid, successive, mitotic division of zygote).

In Vitro Fertilization

In vitro fertilization is a process by which the ovum is fertilized by sperm outside of the body.

In Vitro Fertilization or **IVF** involves uniting sperm and ovum to create embryos in the lab (*"in vitro"*). Once embryos are created, they are placed in the uterus. Thus the embryos are in uterine cavity growing into normal fetuses.

Procedure

The sperms are separated from the semen by laboratory procedure.

The active sperms are combined in a laboratory dish with ova.

About 18 hours after the fertilization procedure, it is possible to grow as embryo.

ICSI (Intracytoplasmic Sperm Injection)

It is another advanced *in vitro* fertilization where the sperms are injected directly into the cytoplasm of the ovum outside the female reproductive system and then transferred into the uterus after the fertilization.

BLASTOCYST (FIG. 1.6)

After fertilization the oocyte divides by mitosis. The cells of the fertilized ovum are called the **blastomeres** and the process of rapid division of cells by mitosis is called the **cleavage**. The cells are all surrounded by zona pellucida. The multiplying cells have the appearance of mulberry and the mass of cells is called the **morula**. As the cell division continues, the morula is propelled slowly into the uterine cavity by peristaltic waves and ciliary beats of columnar cells, of the uterine tube. The secretion of the glandular cells of mucosa provides the fluid vehicle for the transportation and nutrition to the morula.

Between the 5th and 8th day after the fertilization, the morula reaches the uterine cavity (12-16 cell stage). The fluid from the uterine glands of the endometrium and the secretory cells of the uterine tube diffuse through the zona pellucida to accumulate between the blastomeres in small spaces. The cells in the morula simultaneously differentiate into a central mass and peripheral layer. The spaces filled with the uterine fluid enlarge and coalesce to form a single, large cavity. The **central mass of cells**, which is also called the **inner cell mass or embryonic mass**, is pushed to the periphery and is attached to the inner surface of the peripheral layer at one end. The peripheral layer is called the **trophoblast**.

The **embryonic mass** of cells gives rise to **embryo proper** and **trophoblast** gives rise to **coverings of the embryo**.

The embryonic mass is separated from the trophoblast by a cavity except at one end and that end is called the embryonic pole. This cystic mass of cells after fertilization is called the **blastocyst** (Fig. 1.6). The cavity of the blastocyst is called the **blastocele** which contains the uterine fluid for the nourishment of the cells of the blastocyst. This stage is called the **blastula**.

The entry of uterine fluid into the blastocele leads to increase in the size of the blastocyst which in turn leads to thinning and disappearance of the zona pellucida and the blastocyst adheres to the uterine mucosa.

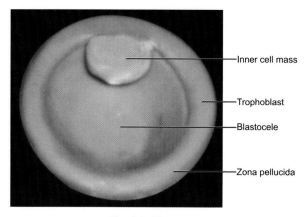

Fig. 1.6: Blastocyst

TROPHOBLAST

The blastomeres in the fertilized ovum keep dividing by mitosis and undergo differentiation into an inner cell mass and an outer layer. The inner cell mass is the embryonic mass which gives rise to the embryo proper and the outer layer is called the trophoblast which gives rise to the coverings of the embryo. This differentiation takes place at the morula stage and reaches the uterine cavity around 4th or 5th day after the fertilization.

The trophoblastic cells form protective and nourishing membranes of the embryonic mass which gives rise to embryo. The trophoblast cells are smaller than the inner cell mass and rapidly divide, lose their cellular configuration and spread rapidly.

The trophoblast dissolves the uterine mucosa and burrows in the stratum compactum of uterine mucosa. This extending trophoblast forms continuous sheet and appears multinucleated because the cells are multiplying before the appearance of the cell membrane. It is called the **syncytiotrophoblast**. The deeper layer of trophoblast cells get back their cell membranes gradually and are called the **cytotrophoblast**.

The trophoblast helps for the implantation of the blastocyst. Due to the irregular growth of the trophoblast the blood vessels are invaded and blood accumulates in small cavities in the trophoblast which are now called the lacunae. Due to the continuous growth of the trophoblast, it grows in the form of finger-like processes called **chorionic villi** and the lacunae between them are called the **intervillous spaces**. They contain the **maternal blood**.

The cytotrophoblast grows and invades through the syncytiotrophoblast and comes in contact with the stratum basalis of the uterine mucosa. It grows horizontally and completely separates the syncytiotrophoblast from the stratum basalis of uterine mucosa. This cytotrophoblastic covering is called **cytotrophoblastic shell**.

In the later part of pregnancy the cytotrophoblast disappears and makes the placental membrane thin to facilitate permeability to substances having smaller molecular weight.

IMPLANTATION (FIG. 1.7)

The blastocyst increases in size due to diffusion of uterine fluid and the **zona pellucida thins out and disappears**. The trophoblast now comes in contact with the epithelium of the endometrium which has reached the thickness of 5 mm. The trophoblast cells secrete a sticky substance which also causes lysis of the tissues whichever it comes in contact. By this process of **adhesion**, **erosion and invasion** the blastocyst gets inserted into the uterine endometrium. This process is called **implantation** (Fig. 1.7).

Erosion of epithelium, stroma, lining epithelium of uterine glands and the endothelium of blood vessels of the endometrium takes place and ultimately the trophoblast, which later becomes chorion, comes in contact with the blood of the uterine wall. This type of contact is called hemochorial.

The Normal Site of Implantation

The normal site of implantation is in the endometrium of the body of the uterus, most frequently in the upper part of the posterior wall near the midline.

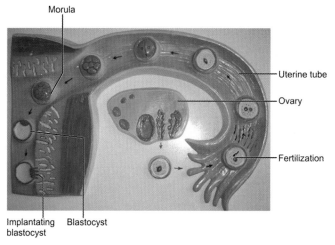

Fig. 1.7: Implantation

Abnormal Sites of Implantation

- Lower part of the uterine mucosa.
- Ampulla of uterine tube.
- Isthmus of uterine tube.
- Infundibulum of uterine tube.
- Ovary.

- Peritoneum of broad ligament.
- Mesentery of intestine.

Sometimes the implantation in the uterine tube or ovary continues up to 6th or 10th week and the tube or the ovary ruptures slowly and developing fetus grows in the abdominal cavity. This is called secondary implantation.

DECIDUA (FIG. 1.8)

The word 'decidua' is derived from deciduous, that means to 'shed off'

It is a **modified endometrium** after the fertilization which is shed off after nine months.

After the fertilization and successful implantation, a hormone secreted by the trophoblast called the chorionic gonadotropin, prolongs the life of the corpus luteum which continues to secrete the progesterone hormone. This hormone increases the vascularity of the endometrium that maintains the life of the endometrium. The stroma thickens, the glands become more secretory and the blood vessels increase. The interglandular tissue contains more leukocytes and crowded with large, round, oval or polygonal **decidual cells**. The decidual cells are modified stromal cells which have accumulated glycogen and lipid in their distended cytoplasm. The cells probably are concerned with nutrition or defensive mechanism.

After the blastocyst is embedded or implanted, distinctive names are applied to different regions of the decidua. The part which covers the conceptus is called the **decidua capsularis**, the part between the conceptus and the uterine muscular wall is called the **decidua basalis** and the

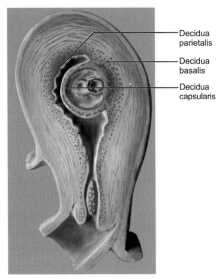

Decidua
parietalis

Decidua
basalis

Decidua
capsularis

Fig. 1.8: Decidua and chorionic sac

part which lines the remainder of the body of the uterus is known as **decidua parietalis** (Fig. 1.9).

Due to the growth of the embryo and the expansion of the cavity of amnion, the decidua capsularis is thinned out, distended and the space between it and deidua parietalis is gradually obliterated. By the third month of gestation, the decidua capsularis and the decidua parietalis are in

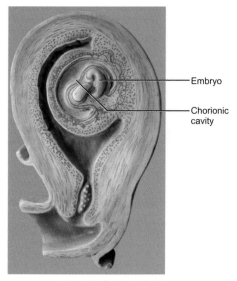

Fig. 1.9: Parts of decidua

contact with each other. Gradually in 5th month both the layers are thinned out and disappear (Fig. 1.10).

The glands in the stratum compactum are obliterated and their epithelium is lost. In the stratum spongiosum, the glands are compressed, appear as oblique slit-like fissures and their epithelium degenerates. The epithelium of the glands is retained; they are cuboidal in nature in the basal zone.

Fig. 1.10: Fate of decidua

TWIN PREGNANCY

One of the two fetuses developed within the uterus at the same time from the same pregnancy, is called twin pregnancy (Fig. 1.11).

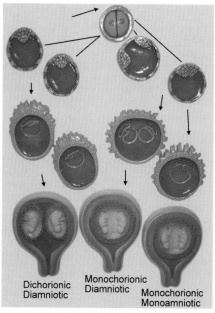

Dichorionic Diamniotic

Monochorionic Diamniotic

Monochorionic Monoamniotic

Fig. 1.11: Twinning

2

Fetal Membranes

Fetal membranes include the following:
- Amnion
- Yolk sac
- Extraembryonic celom
- Allantois
- Connecting stalk/umbilical cord
- Chorion
- Placenta

AMNION

Amnion is a thin, tough, fluid-filled, membranous sac that surrounds the embryo and the fetus. It is attached to the margins of the embryonic disk.

The amniotic sac contains the amniotic fluid, which helps to keep the baby warm and keeps the growing fetal parts from fusing together.

To begin with, the amniotic fluid is a transudate of the maternal plasma and becomes more like the fetal fluids only in the presence of the fetus. When the fetus is removed, the fluid remains similar to maternal plasma.

The origin of amnion is controversial:

Whether it develops from embryoblast as amniogenic cells and these cells separate out from epiblast and surround the space called amnion or whether the trophoblastic cells differentiate into amniogenic cells or due to delamination of embryoblast and trophoblast from each other to enclose a slit which becomes amnion.

As amnion enlarges, it gradually obliterates the chorionic cavity and forms epithelial covering of umbilical cord and placenta.

Amniotic fluid plays major role in fetal growth and development.

The fluid is initially secreted by amniotic cells but later most of the fluid is derived from maternal tissue and by diffusion across amniochorionic membrane from decidua parietalis. This is followed by diffusion of fluid through chorionic plate from blood in the intervillous space of placenta.

Before keratinization of the skin the major pathway for diffusion of water and solutes from fetus to amniotic cavity is through skin, thus the fluid is similar to tissue fluid.

The other sources are:
- Respiratory tract.
- Excreted urine by the fetus (about 300-400 ml).
  By late pregnancy half a liter of urine is added daily.
  The volume of amniotic fluid—30 ml at 10 wk.
  350 ml at 20 wk and about a liter by full-term.

Composition of amniotic fluid:
- 99% water
- 1% organic and inorganic substances.
  Half of the organic constituents are proteins, the remaining half consist of carbohydrate, fat, enzyme, hormones, pigments and desquamated epithelial cells.

YOLK SAC (FIGS 2.1, 2.2A AND B)

Formation and Boundaries

It is the cavity of the blastula (blastocele), after the formation of the first germ layer called the hypoblast. It begins as primordial or **primary yolk sac** (Fig. 2.1).

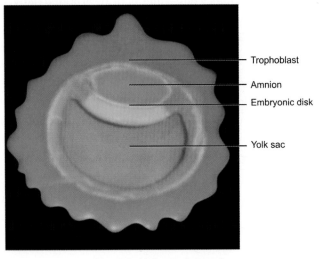

Fig. 2.1: Yolk sac (primary)

The primary yolk sac becomes smaller when the extraembryonic mesoderm is formed and is called **secondary yolk sac**.

After the formation of the extraembryonic celom the secondary yolk sac is further reduced in size. Part of the secondary yolk sac is pinched into the midgut when the foldings of the embryo take place and there is formation of gut tube. It is then called the **definitive yolk sac**. The narrow, compressed part of the yolk sac which connects the distal small yolk sac and the midgut is termed the **vitellointestinal duct (Figs 2.2 A to C)**.

Functions

- Nutrition
- Primordial germ cells formation
- Production of blood in the wall of the yolk sac
- Takes part in the formation of primitive gut tube.

Fate of Yolk Sac

By 10th week it is very small and lies between amniotic and chorionic sacs.

It atrophies as pregnancy advances

In abnormal conditions the yolk stalk may persist and give rise to Meckel's diverticulum.

EXTRAEMBRYONIC CELOM (FIGS 2.2 A TO C)

It is the space in the extraembryonic mesoderm. Also called chorionic sac.

Number of small, isolated spaces appear in the mesoderm, increase in size and all of them rapidly fuse to form a large, isolated cavity – the **extraembryonic celom**.

This fluid-filled cavity gradually surrounds the embryonic disk, yolk sac and amnion except where they are attached to the chorion by connecting stalk. Due to the increasing size of the EEC, the primary yolk sac decreases in size and a smaller secondary yolk sac is formed (Figs 2.2A to C).

The cavity splits the extraembryonic mesoderm into an outer layer called Somatopleuric Mesoderm and an inner layer called the Splanchnopleuric Mesoderm.

Extraembryonic splanchnopleuric mesoderm surrounds the yolk sac.

The extraembryonic somatopleuric mesoderm lines the trophoblast and amnion. This, along with trophoblast forms 'Chorion'. Thus the chorion forms the wall of the chorionic

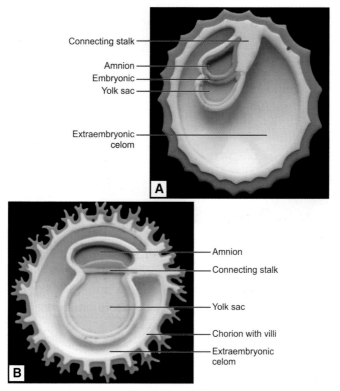

Figs 2.2A and B

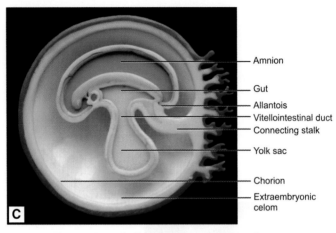

Figs 2.2A to C: Extraembryonic celom, yolk sac,
amniotic cavity and connecting stalk

sac within which embryo, amnion and yolk sac are
suspended by connecting stalk. Therefore extraembryonic
celom is also called chorionic cavity.

Transvaginal ultrasound (endovaginal sonography) is
used to measure chorionic sac diameter.

Gradually amnion grows in size and expands into
extraembryonic celom.

ALLANTOIS (FIGS 2.3A TO C)

Formation

On 16th day, a sausage-shaped diverticulum from the caudal wall of yolk sac appears which extends into the connecting stalk. This is called allantois.

It remains very small in human embryos.

During the 2nd month the extraembryonic part of the allantois degenerates and a portion extends into connecting stalk (Figs 2.3A to C).

The allantois gets incorporated into ventral wall of hindgut after the foldings of the embryo.

The postallantoic part of the hindgut forms cloaca.

Functions

– Blood formation between 3rd and 5th week of IUL.
– Development of umbilical vessels.
– In reptiles, birds and mammals it has respiratory function and acts as reservoir of urine.

Fate

It extends between urinary bladder and umbilicus. As the bladder enlarges the allantois involutes to form **urachus**. Later on the urachus degenerates to form **median umbilical ligament**.

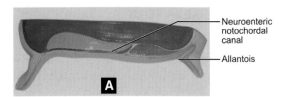

Neuroenteric notochordal canal

Allantois

A

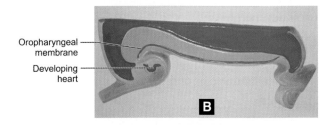

Oropharyngeal membrane

Developing heart

B

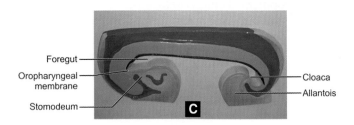

Foregut

Oropharyngeal membrane

Stomodeum

Cloaca

Allantois

C

Figs 2.3A to C: Allantois

CONNECTING STALK

On day 14, the embryonic disk with its amnion and yolk sac becomes suspended in the chorionic cavity by a thick layer of mesoderm, which elongates to form the connecting stalk (Fig. 2.4). The mesoderm consists of both parietal and visceral extraembryonic mesoderm. It is gradually lined by the extraembryonic celomic epithelium. It elongates and forms umbilical cord that connects the fetus to the mother.

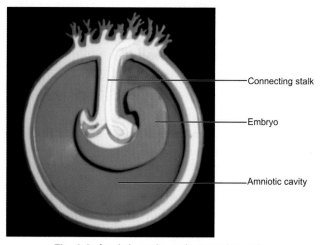

Connecting stalk

Embryo

Amniotic cavity

Fig. 2.4: Amniotic cavity and connecting stalk

47

UMBILICAL CORD

It consists of:

An outer covering of amniotic membrane.

An interior mass of mesoderm derived from:

a. Somatopleuric extraembryonic mesoderm covering amniotic folds.
b. Splanchnopleuric extraembryonic mesoderm of yolk sac.
c. Splanchnopleuric extraembryonic mesoderm of allantois and umbilical vessels.
d. Vitellointestinal duct and allantois.
e. Vitelline and umbilical vessels.
f. Remains of extraembryonic celom.

The amniotic membrane is made up of epithelial cells which are simple squamous or cuboidal cells.

The mesoderm core from different sources fuses and gradually transforms into a viscid, mucoid connective tissue which is called 'Wharton's jelly'.

Wharton's jelly consists of:

1. Widely spaced fibroblasts (with branching processes) separated by extensive intercellular space.

2. Copious matrix—Consisting of delicate, three dimensional meshwork of fine collagen fibers.
3. Ground substance (with a variety of hydrated mucopolysaccharides).

Yolk sac and allantois slowly degenerate and disappear. Only interrupted cords of cells may be seen.

Vitelline vessels atrophy in the region of the umbilical cord.

Umbilical vessels are – 2 arteries and 2 veins – right and left.

The right umbilical vein disappears in the early months of gestation.

The vessels are twisted—in right or left handed cylindrical helix (Fig. 2.5).

The umbilical cord begins to form about five weeks after conception.

It becomes progressively longer until about 28 weeks of pregnancy, reaching an average length of 22 inches.

As it gets longer, the cord generally twists around itself and becomes coiled.

The number of twists or turns may exceed 300.

They are due to unequal growth of vessels and torsion forces imposed by fetal movements.

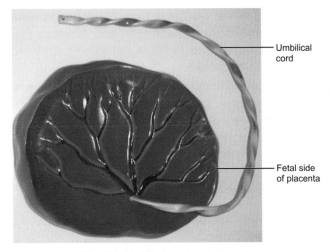

Fig. 2.5: Umbilical cord and placenta

The arteries are medium-sized, muscular arteries.

The external layers of tunica media of the arteries pursue oblique/spiral course.

Contraction produces shortening of vessels and thickening of tunica media. The pulsations of the umbilical arteries assist venous return to fetus in umbilical vein. This leads to folding/bulging of tunica intima and act as valves during pulsation.

These foldings are called 'Valves of hoboken'.

Cord exhibits two types of knots, namely.

- **True knots**—due to fetal movements.
- **False knots**—due to sharp variation in contour caused by local accumulation of mass of Wharton's jelly.

CHORION

Formation

The trophoblast and the somatopleuric extraembryonic mesoderm are together called the 'chorion'.

It grows towards the basal plate (Decidua basalis) to form placenta.

Parts

Consists of the thick part called the **chorion frondosum** which is nearer the basal plate and the thin, smooth part called **chorion leavae** which is at the region of decidua capsularis.The chorion frondosum consists of thickly arranged branching, finger-like projections called the **chorionic villi**.

Chorionic villi are of three types depending upon the content in the wall:

a. **Primary chorionic villi** – Contain only cytotrophoblast and syncytiotrophoblast (Fig. 2.6).

b. **Secondary chorionic villi** – Contain outer syncytiotrophoblast, inner cytotrophoblast and a central core of extraembryonic, somatopleuric mesoderm (Fig. 2.7).

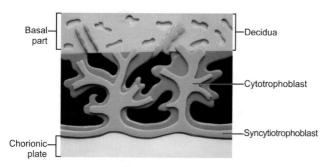

Fig. 2.6: Primary chorionic villi

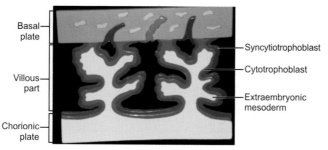

Fig. 2.7: Secondary chorionic villi

c. **Tertiary chorionic villi** – Outer syncytiotrophoblast, deeper cytotrophoblast, inner extraembryonic meso-derm and the fetal capillaries (Fig. 2.8).

53

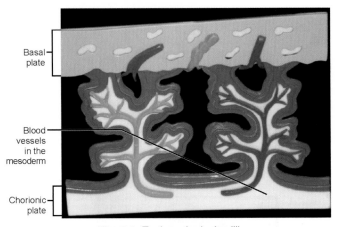

Fig. 2.8: Tertiary chorionic villi

Depending upon the attachment of the villi to the basal plate, the chorionic villi are classified into:
a. Anchoring villi
b. Floating villi.

According to the branching pattern:
a. Truncus chorii
b. Rami chorii
c. Ramuli chorii.

Depending upon the thickness and amount of the chorion and villi the chorion consist of:

a. Chorion frondosum (thick part)

b. Chorion leavae (thin, smooth part).

PLACENTA (FIGS 2.8 AND 2.9)

The word 'Placenta' is derived from 'plakuos' a Latin word which means 'Flat Cake'.

It is the organ, characteristic of true mammals during pregnancy, joining mother and fetus, provides nutrition, hormones and helps in removal of waste.

Full-term Placenta

Definition

It is a fetomaternal tissue which is discoidal, deciduate, hemochorial, chorioallantoic, initially labyrinthine, later a villous organ.

Shape is disk-like (discoidal)

Is shed off (deciduate)

Fetomaternal connection is chorion and blood (hemochorial).

It is vascularized by allantoic vessels of body stalk (chorioallantoic).

The trophoblast has interconnecting tunnels (labyrinth), which become finger-like later (villous).

Measurements of Full-term Placenta (See Fig. 2.5)

Diameter = 185 mm (range is 150-200 mm).

Weight = 500 gm (range 200-800 gm).

Thickness = 23 mm (range = 10-40 mm).

Surface area = 30,000 sqmm

Thickest at the center

Has two surfaces—fetal and maternal.

Fetal surface is smooth, contains blood vessels and is covered by amniotic membrane. Umbilical cord is attached in the middle (Fig. 2.9).

Maternal surface is finely granular and mapped into 15 to 20 irregular, polyhedral lobes separated by fissures or grooves called 'Cotyledons' or 'Uterine caruncles' (each cotyledon is called as lobe by obstetricians).

The grooves are the sites of placental septa.

Fetal unit—Area supplied by major umbilical vessels—along with corresponding maternal tissue is called 'placentome'.

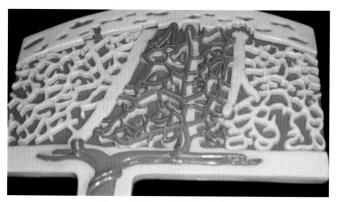

Fig. 2.9: Placental circulation

The fetal cotyledon corresponds to major villous stem and its branches.

Early in pregnancy the chorion bears 800 to 1000 such stems but as gestation advances, due to disappearance and fusion of remaining persisting stems, the number is reduced to 60 which are distributed between 15-20 lobes.

Therefore, each lobe contains 2-4 major villus stems.

These cells are rich in organelles and inclusion bodies, indicating highly active cells.

As gestation advances, the endothelial cells and syncytiotrophoblastic cells become thin and elongated.

3

Development in Third Week of Life

The following structures are formed in third week:
- Primitive streak
- Notochord
- Neural tube
- Neural crest
- Somites
- Intraembryonic celom
- Blood vessels and blood of the fetus
- Completion of chorionic villi development
- Prochordal plate and cloacal membrane.

PRIMITIVE STREAK

The **primitive streak** is a **linear band of thickened epiblast** that first appears at the **caudal** end of the embryo and **grows cranially**. At the cranial end its cells proliferate to form the **primitive knot (primitive node)**. With the appearance of the primitive streak it is possible to distinguish cranial (primitive knot) and caudal (primitive streak) ends of the embryo.

Primitive knot and the streak are formed in the following way:

1. **Cell proliferation**—Causes heaping up of the cells of ectoderm caudal to the prochordal plate and epiblast is the source of a new layer of cells.

2. **Cell migration** by amoeboid movement—The cells insinuate themselves between the epiblast and hypoblast.

3. **Cell determination**—The cells arising from the primitive streak are determined to give rise to different rudiments.

The cells of the primitive streak which insinuate between the epiblast and hypoblast extend all over the embryo except the prochordal plate cranially and cloacal membrane caudally. Thus, a third layer is formed which is called the **intraembryonic mesoderm** (**secondary mesoderm**) (Figs 3.1 and 3.2). By this process cells of the epiblast are translocated to new positions in the embryo, producing the **three primary germ layers**. Formation of the secondary or intraembryonic mesoderm is called the **Gastrulation**.

- Gastrulation is a crucial time in the development of multicellular animals. During gastrulation, several important things are accomplished:
  - The three primary germ layers are established.

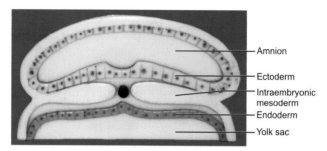

Fig. 3.1: Intraembryonic mesoderm

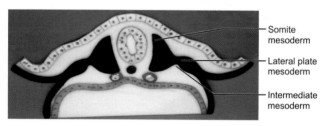

Fig. 3.2: Fate of intraembryonic mesoderm

   – The basic body plan is established, including the physical construction of the rudimentary primary body axes.

As a result of the movements of cells in gastrulation, cells are brought into new positions, allowing them to interact with cells that were initially not near them. This paves the way for inductive interactions, which are the hallmark of neurulation and organogenesis.

Changes Consequent to Gastrulation

With the process of gastrulation the following changes have occurred:

1. The embryo becomes a **trilaminar embryo** but is still in the form of a flat disk.
2. Epiblast and hypoblast are now known as **ectoderm** and **endoderm** respectively.
3. Mesoderm does not extend between epiblast and hypoblast at the prochordal plate (buccopharyngeal membrane) and cloacal membrane.
4. Intraembryonic mesoderm merges with the extra-embryonic mesoderm at the periphery of the embryonic disk.

DEVELOPMENT OF NOTOCHORD

It is the **fore runner** of **axial skeleton**.

It is the primary inductor in the early embryo, induces overlying ectoderm to form neural plate.

Notochord develops as follows:

A solid cord of cells grow from the cranial end of the primitive knot towards the prochordal plate. This process is called the **head process** or **notochordal process**.

This process pierces the endoderm just caudal to the prochordal plate and comes in contact with the yolk sac.

The notochordal process now undergoes canalization from the continuation of a depression, called **blastopore**, at the center of the primitive knot.

The canal is called **notochordal canal (neuroenteric canal)**, which connects the amniotic cavity with the yolk sac.

This connection helps in nourishing the endoderm and its derivatives till the blood supply to the deeper structures is established.

The ventral aspect of the notochordal canal breaks at multiple points and the canal communicates with yolk sac at multiple levels.

Gradually the floor (ventral wall) of the notochordal canal disappears and **notochordal plate** followed by notochordal **groove** is formed.

The ends of the notochordal groove separate from the endoderm and join to form the **notochord**.

The separated ends of the endoderm now join to form a continuous endodermal layer. The connection between the amniotic cavity and the yolk sac will be closed as the notochord is formed.

The notochord degenerates to form nucleus pulposus of intervertebral disk, apical ligament of dens, basiocciput and basisphenoid in the adult life.

FORMATION OF NEURAL TUBE

The neuroectoderm is derived from the epiblast and is induced by the underlying notochord during the third week. It extends along the axis of the embryo just dorsal to the notochord. It is called the **neural plate** (Fig. 3.3A).

Late in the third week the neural plate begins to fold. It is first converted into a **neural groove** (Fig. 3.3B).

The neural groove deepens and eventually forms a **neural tube** (Figs 3.3C and D).

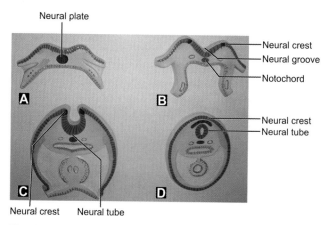

Figs 3.3A to D: Formation of neural plate, groove, tube and neural crest

Two masses of ectoderm along the edges of the neural plate form the **neural crest** (Figs 3.3B to D).

Initially the neural crest separates neuroectoderm from skin ectoderm.

As folding of the neural tube occurs, the neural crest cells detach from the ectoderm and form clusters that migrate into the mesoderm (Fig. 3.3D).

This process of formation of neural tube and neural crest is called the **Neurulation** (Fig. 3.3).

The neural tube elongates, undergoes folding called the flexures and cranial part dilates at three parts separated by constrictions which are called the **primary brain vesicles** (Fig. 3.4A). These primary brain vesicles undergo further subdivisions into **secondary brain vesicles** which give rise to various parts of the **brain**. Rest of the neural tube caudal to the brain vesicles forms the **spinal cord** (Fig. 3.4B).

The primary inductor for the formation of neural ectoderm is the notochord.

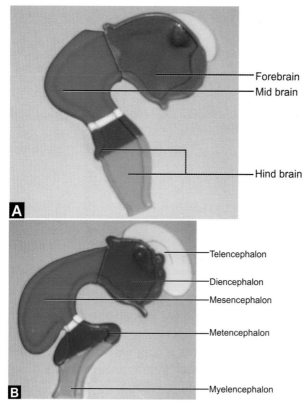

Figs 3.4A and B: Neural tube

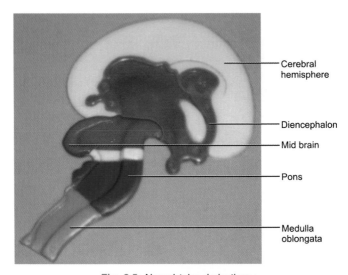

Fig. 3.5: Neural tube derivatives

NEURAL CREST

The neural crest is a specialized subpopulation of ectodermally derived cells originating at the border between the neural plate and future epidermis in the dorsal aspect of all vertebrate embryos (See Fig. 3.3D).

Neural crest cells delaminate and migrate throughout the body and contribute to an impressive variety of tissues ranging from neurons and glia to connective tissue components of the head.

The cardiac neural crest migrating into the heart region is responsible for forming the entire musculoconnective tissue wall of the large arteries emerging from the heart, the membranous portion of the ventricular septum, and the septum between the aorta and pulmonary artery.

In addition, the neural crest contributes to the parathyroid, thyroid, and thymus glands which develop from the pharyngeal apparatus (Fig. 3.5).

Neural crest derivatives are:
- Melanocytes
- Ganglia of dorsal root of common spinal nerve
- Schwann cells
- Adrenal medulla
- Sensory ganglia of V, VII, VIII, IX, X cranial nerves
- Autonomic ganglia
- Enamel of tooth
- Parafollicular cells of thyroid
- Cells of pia mater
- Cells of arachnoid mater

- Bipolar cells of sensory epithelium
- Multipolar neurons
- Cartilage and connective tissue of the third, fourth, and sixth pharyngeal arches.

SOMITES

As the notochord and neural tube develop, the mesoderm along the lateral side forms longitudinal columns called **paraxial mesoderm**. These columns divide into paired cubical bodies called **somites** (Fig. 3.6). The somites develop in pairs; the first pair develops near the cranial end of the notochord around the end of the third week. Additional pairs of somites develop in a caudal direction from day 20 to 30. Totally about 44 pairs of somites appear in the embryo. The number of somites is sometimes used as a criterion for determining the age of embryo. The somites give rise to most of the axial skeleton like vertebral column, ribs, sternum, and skull base and associated musculature, as well as to the adjacent dermis.

Somite is divided into two parts:

a. The ventromedial part—**Sclerotome**. It contains a "cavity" of loose cells. Cells from the sclerotome migrate medially to surround the notochord and neural tube and form the axial skeleton.

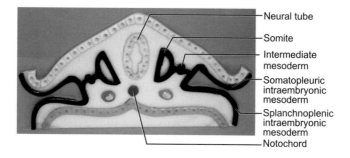

Neural tube

Somite

Intermediate mesoderm

Somatopleuric intraembryonic mesoderm

Splanchnoplenic intraembryonic mesoderm

Notochord

Fig. 3.6: Somites

b. The dorsolateral part—**dermomyotome**. Cells from the
    dermomyotome migrate laterally and, as its name
    implies, gives rise to: (i) skeletal muscle and (ii) the
    dermis of the skin. The myotome contains a cavity in
    the early stages called the myocele which obliterates
    immediately.

The concept of the myotome in gross anatomy is an
embryological concept. Each anatomical myotome is
derived from the embryological dermomyotome that is
innervated by a segmental nerve and forms a group of
skeletal muscle and the dermis of the corresponding
segment of skin.

EMBRYONIC FOLDING

Folding occurs by differential growth of tissues. Neural ectoderm grows faster than the surrounding skin ectoderm and consequently folds to form a neural tube. Similarly, skin ectoderm grows faster than the underlying mesoderm and endoderm, and this differential growth causes folding of the trilaminar disk and gives shape to the embryo.

This period is also called **period of organogenesis**.

It is the critical period of embryo (Figs 3.7 and 3.8).

All major external and internal structures are established during this period.

The main organ systems have begun to develop but function of most of them is minimal except cardiovascular system.

As the tissue and organs form, the shape of the embryo changes and by the end of 8th week, it has a distinct human appearance.

Phases of Embryonic Development

- First phase—Growth
- Second phase—Morphogenesis
- Third phase—Differentiation.

Primitive pericardial cavity

Fig. 3.7: Sagittal section of embryonic disk before folding

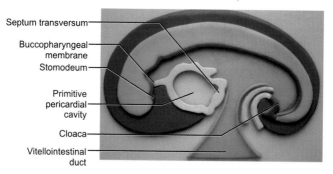

Septum transversum

Buccopharyngeal membrane

Stomodeum

Primitive pericardial cavity

Cloaca

Vitellointestinal duct

Fig. 3.8: Sagittal section of embryonic disk showing foldings of the embryo

Folding occurs mainly at the edges of the embryonic disk and forms three main folds (Fig. 3.8):

1. Head fold
2. Tail fold
3. Lateral folds—convert the embryo into a tubular structure.

These are not three separate folds but occur simultaneously and merge into one another.

The notochord, neural tube and somites stiffen the dorsal axis of the embryo.

As a result of the formation of the **head fold**:

a. The **foregut** is formed by folding of the endoderm
b. The **stomodeum** is an invagination of ectoderm, and has the buccopharyngeal membrane separating it from the foregut: It opens into the amniotic cavity.
c. The **pericardial cavity** and cardiogenic mesoderm are shifted to the ventral aspect of the embryo and lie **ventral** to the foregut.
d. The part of the transverse mesoderm between the pericardial cavity and the yolk sac is the **septum transversum**. In it the liver will develop (Fig. 3.8).

e. The **amniotic cavity** extends ventral to the cranial end of the embryo.
f. The **yolk sac** is constricted from the cranial aspect.

As a result of the formation of the **tail fold**:

a. The hindgut is formed.
b. The connecting stalk is shifted ventrally.
c. The allantoic diverticulum is shifted ventrally. It is an invagination of hindgut endoderm into the yolk sac.
d. The amniotic cavity extends ventral to the caudal end of the embryo.
e. The yolk sac is constricted from the caudal end.

Transverse or Lateral Folding of the Embryo

a. Converts the endoderm into a primitive gut tube.
b. The intraembryonic celom surrounds the gut tube.
c. The communication between the intra- and extra-embryonic celoms becomes constricted and eventually obliterated.

Important changes occur in the embryonic cavities as a consequence of folding:

1. The **amniotic cavity** surrounds the embryo completely on all aspects and becomes the predominant cavity. It enlarges progressively (Fig. 3.8).

2. The **yolk sac** becomes constricted on all sides, and becomes a small sac connected to the midgut by a narrow vitelline duct. It becomes progressively smaller.

3. The **extraembryonic celom** is gradually obliterated by the expanding amnion and eventually disappears completely.

DERIVATIVES OF GERM LAYERS

Ectoderm

Central nervous system.
Peripheral nervous system.
Sensory epithelia of the eye, ear, and nose.
Mammary glands.
Enamel of teeth.
Epidermis and epidermal derivatives of the integumentary system, including the hair follicles, nails, and the glands that communicate with the skin surface.
The lining of vestibule of oral cavity.
Parotid salivary glands.
Anal canal below pectinate line.
Portions of the skull, pharyngeal clefts.

Anterior pituitary.
Medulla of adrenal glands.
Bone of pharyngeal (branchial) arch origin.
Ciliaries muscle of eyeball.
Arrectores pilorum muscle of skin.
Melanocytes.

Endoderm

Epithelial lining of the gastrointestinal tract.
Epithelial lining respiratory tract.
Epithelial lining of the tympanic cavity.
Epithelial lining of mastoid antrum.
Epithelial lining of auditory tube.
Epithelial lining of the urinary bladder and most of the urethra.
Liver.
Pancreas.
Parathyroid glands.
Thymus.
Thyroid.
Tonsils.

Mesoderm

Blood.
Bone.
Cartilage.
Connective tissue.
Cortex of adrenal glands.
Genital ducts (mesonephric and paramesonephric ducts).
Heart.
Kidneys.
Lymphatic vessels.
Ovaries.
Testis.
Serous membranes of the body cavities (pericardium, pleura, and peritoneum).
Spleen.
Striated and smooth muscles.
Dermis of the skin.

4

Developments in Fourth Week and Onwards

BRANCHIAL APPARATUS

Around 4th week of gestation, as the neural crest cells migrate into future head region, the pharyngeal apparatus begins to develop.

This apparatus develops in the future pharynx therefore called 'pharyngeal apparatus' (Fig. 4.1).

It appears in the region of the visceral mesoderm around the cranial part of foregut surrounded by endoderm inside and ectoderm outside, therefore it is also called 'visceral apparatus'.

This region of the foregut represents the gills of the fish, therefore called 'branchial apparatus' (branchia, in Greek, means Gills).

Between the stomodeum and developing heart, in the visceral mesoderm of the foregut, horizontal thickenings appear, (initially the visceral mesoderm later by neural crest).

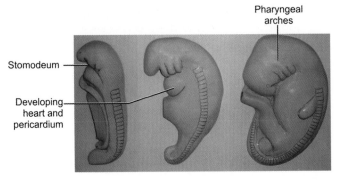

Fig. 4.1: Pharyngeal apparatus

These horizontally thickened bars grow forwards/ anteriorly and fuse with each other in the anterior midline to form **pharyngeal/branchial arch**.

Thus, six such arches appear which displace the heart downwards. Later the **5th arch disappears**.

Each arch has a mesodermal core covered externally by ectoderm and internally by endoderm.

Internally, the endoderm grows between adjacent arches in the form of diverticulum and is called **pharyngeal/ branchial pouch**. There are 4 pouches and 5th pouch is rudimentary.

Externally, the ectoderm, presents shallow depressions between the adjacent arches, which are termed the **pharyngeal/branchial clefts**. 4 clefts are present.

Between the arches the ectoderm and endoderm are separated by a thin layer of mesoderm which is called **branchial plate** or **pharyngeal membrane**.

In the fish, these branchial plates rupture to form gill slits. In the human, they do not rupture and are soon invaded by the mesoderm of the neighboring arches.

Pharyngeal Arches

Derivatives of **pharyngeal arches (Fig. 4.2)**:

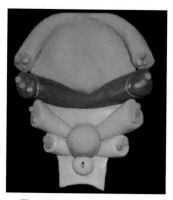

Fig. 4.2: Pharyngeal arches

Arch	Nerve	Derivatives in the adult	
		Muscle	Skeletal derivatives
First (Mandibular)	**Posttrematic** – Mandibular branch of Vth cr. nerve. **Pretrematic** – Chorda tympani branch of VIIth cr. nerve	Muscles of mastication, Tensor palatini, tensor tympani, mylohyoid, anterior belly of digastric.	Malleus, incus, mandible, anterior ligament of malleus. Sphenomandibular ligament.
Second (Hyoid)	**Post trematic** – VIIth cr. nerve	All the muscles of facial expression, auricular muscles, scalp muscles, stapedius, stylohyoid, posterior belly of digastric.	Stapes, styloid process of temporal bone, lesser cornu and upper part of the body of hyoid bone. Stylohyoid ligament.
Third	**Posttrematic** – IXth Nerve	Stylopharyngeus muscle, constrictor muscles of pharynx (?)	Greater cornu and lower part of body of hyoid bone.
Fourth	**Posttrematic** – Superior laryngeal branch of vagus (Xth cr.) nerve.	Muscles of soft palate except tensor palatini, muscles of pharynx except stylopharyngeus, cricothyroid.	Laryngeal cartilages.
Sixth	**Posttrematic** Recurrent laryngeal nerve	Intrinsic muscles of larynx except cricothyroid	Laryngeal cartilages.

Pharyngeal Pouches

The internal surfaces of pharyngeal arches are lined by the endoderm of foregut.

Endodermal diverticula between pharyngeal arches are called pharyngeal pouches (Fig. 4.3).

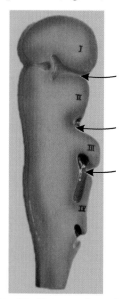

Fig. 4.3: Pharyngeal pouches

Derivatives of Pharyngeal Pouches

Ist. **Pharyngeal pouch**: Ventral part—Tongue and pharynx.

Dorsal parts of I and II pouches- tubotympanic recess—Forms middle ear, mastoid antrum and auditory tube.

IInd. **Pharyngeal pouch**: Ventral part—Tonsil.

Dorsal part – Contributes to tubotympanic recess—Forms middle ear, mastoid antrum and auditory tube.

IIIrd. **Pharyngeal pouch**: Ventral part—Thymus.

Dorsal part—Inferior parathyroid.

IVth. **Pharyngeal pouch**: Ventral part—Thyroid (lateral thyroids).

Dorsal part—Superior parathyroid.

(Rudimentary V pouch gives rise to **ultimobranchial body** or caudal pharyngeal complex. Neural crest cells migrate into this body and give rise to parafollicular cells of thyroid gland).

Pharyngeal Clefts

1st cleft forms external auditory meatus. The epithelial lining of the floor of the 1st cleft forms the cuticular layer of tympanic membrane (Fig. 4.4).

From **2nd cleft** onwards, the clefts and arches are covered by the outgrowth of the second arch which overhangs 2nd,

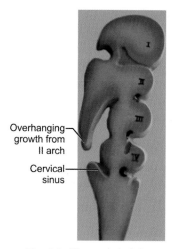

Overhanging growth from II arch

Cervical sinus

Fig. 4.4: Pharyngeal clefts

3rd and 4th clefts and finally fuses with the outer aspects of the remaining arches. Thus, the space between the overhanging second arch and 2nd, 3rd and 4th clefts form **cervical sinus**. In further development the sinus disappears due to the fusion of the overhanging mesoderm of second arch.

Persistent cervical sinus and rupture of pharyngeal membrane leads to communication between pharynx with exterior called **branchial fistula**. The fistula is found along the lateral wall of neck, anterior to sternocleidomastoid muscle. Nonobliteration of cervical sinus results in **cervical/ branchial cyst** which is closed at both ends. The cyst may not be visible at birth but later may be seen as an enlargement along the anterior border of sternocleidomastoid muscle. Cysts can be single or multiple.

DEVELOPMENT OF FACE

There are five primordia or processes involved in the development of face (Figs 4.5A and B).

They are: **Frontonasal**, a pair of **maxillary** and a pair of **mandibular** prominences. All these are around the stomodeum (future mouth).

In the frontonasal process, on either side of the midline, above the stomodeum, there will be ectodermal thickening called the **olfactory or nasal placode**. Deep to the olfactory placode, there is mesodermal thickening at the circumference of the placode which causes elevation along the margins of the placode. The raised margins are called the nasal processes; they are **medial nasal process** and **lateral nasal process**. These nasal processes surround a pit called the **nasal pit**. The nasal pit is floored by olfactory placode. The nasal pit communicates with the

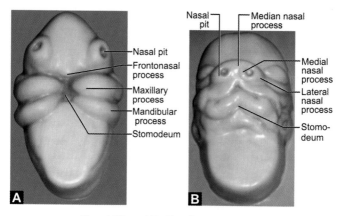

Figs 4.5A and B: Development of face

stomodeum on the inferior aspect. Maxillary prominences that arise from the cranial part of the first arch grow towards midline compressing medial nasal prominences, thereby the cleft between the maxillary and medial nasal **prominences** is lost and they fuse. The maxillary processes grow towards **median nasal process** or **globular process** or simply the frontonasal process, which is between the two medial nasal processes, fusing to form the cheek and the upper lip.

Hence formation of upper lip is by fusion of median nasal prominence and maxillary prominences. The median nasal prominence gives rise to **philtrum** of upper lip.

Lower lip and jaw are formed by mandibular prominences as they merge across the midline.

Maxillary and lateral nasal prominences initially are separated by **nasolacrimal groove**.

Ectoderm in the groove forms a solid cord of cells which later gets canalized and forms **nasolacrimal duct**. Upper end of the duct gets expanded to form **lacrimal sac**.

Nose Development

Frontonasal prominence—Bridge of nose
Medial nasal prominence—Crest and tip of nose and septum of nose
Lateral nasal prominence—Ala of nose.

The muscles of the face (muscles of facial expression) are derived from IInd visceral arch. Therefore, the nerve supply to the muscles of facial expression is by facial nerve and sensory supply is from various branches of trigeminal nerve.

External ears develop from first cleft and auditory hillocks, developing on either side of 1st cleft.

Cleft Lip of Upper Lip

Caused due to failure of fusion of maxillary process with the frontonasal/median nasal/globular process.

It can be unilateral or bilateral.

Unilateral can be on the right side or left side (Figs 4.6 and 4.7).

This anomaly may be associated with cleft palate.

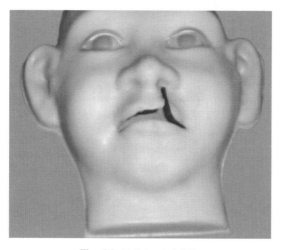

Fig. 4.6: Unilateral cleft lip

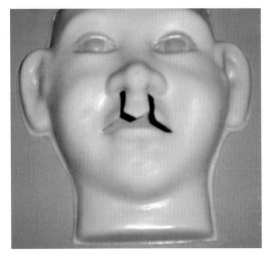

Fig. 4.7: Bilateral cleft lip

Hare Lip

Failure of fusion of mandibular processes—It is in the midline (Fig. 4.8).

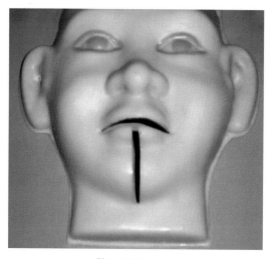

Fig. 4.8: Hare lip

Oblique Facial Cleft

Caused due to nonfusion of maxillary process with the lateral nasal process. It is usually associated with absence of nasolacrimal duct.

It can be unilateral or bilateral.

Unilateral can be on the right side or left side (Fig. 4.9).

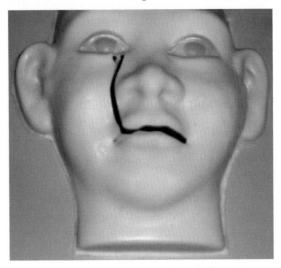

Fig. 4.9: Oblique cleft of the face (unilateral)

DEVELOPMENT OF PALATE (FIG. 4.10)

Primitive or primary palate is formed from intermaxillary segment of frontonasal process. This is adult premaxilla or os incisivum.

Definitive or secondary palate is formed by the fusion of palatal processes or palatal shelves of maxillary prominences with the primitive or primary palate.

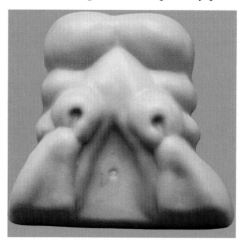

Fig. 4.10: Development of palate

The palatal shelves change from a vertical to horizontal position and fuse with each other and with the primitive palate, to form secondary palate.

The tongue migrates away from the shelves in an antero-inferior direction for palatal fusion to occur.

The anterior two-thirds of the secondary palate undergo ossification and form hard palate.

The posterior parts of the palatine shelves do not ossify and form soft palate.

Palatal Anomalies (Fig. 4.11)

Cleft Palate

Anterior cleft palate is lateral to median plane due to the premaxilla in the midline. Therefore, it is either unilateral or bilateral. It is caused due to nonunion of premaxilla with the palatine shelf of the maxillary process. It may be associated with cleft lip.

The **posterior cleft palate** is always in the midline. It can vary in its extent. It may lead to bifid uvula.

There may be nonunion of palatal shelves and the premaxilla leading to complete cleft palate which is like the letter 'Y'.

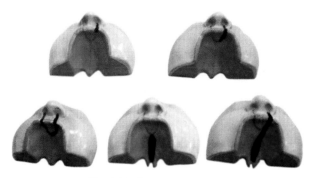

Fig. 4.11: Anomalies of palate

DEVELOPMENT OF TONGUE

The tongue develops in the posterior surface of the anterior part of the pharyngeal apparatus.

On the internal surface of first arch there will be **lingual swellings**, one on each side of the midline.

Between first and second arches, in the midline there is another swelling called the **tuberculum impar**.

Caudally, between third and fourth arches, in the midline, another thickening develops called the **hypobranchial eminence** which consists of cranial and caudal parts.

The lingual swellings grow and fuse with each other.

The cranial part of the hypobranchial eminence **grows** cranially, over the tuberculum impar to join with the lingual swellings. The junction or line of fusion is indicated by **sulcus terminalis**. The tuberculum impar thus gets submerged.

The endoderm covering these swellings give rise to mucous membrane of the tongue. The mesodermal thickenings give rise to connective tissue and muscles of the tongue. The muscles are also developed from **occipital myotomes**.

Due to the **overgrowth** of the hypobranchial eminence on the second pharyngeal arch to fuse with the lingual swellings, the endoderm covering the third arch **spills over** and lies in front of the sulcus terminalis in which the circumvallate papillae develop.

Therefore, the nerve supply is correlated to the development as follows:

Presulcal part develops from lingual swellings of first arch. Hence, general sensory supply is by **lingual branch** of mandibular nerve (posttrematic nerve of Ist arch).

Special sensory supply for the taste buds in the fungiform papillae and secretomotor nerves for the lingual glands is by **chorda tympani** nerve (pretrematic nerve of Ist arch) and for circumvallate papillae, by glossopharyngeal nerve (posttrematic nerve of third arch).

The **postsulcal part** of the tongue is supplied by **Glossopharyngeal nerve** (posttrematic nerve of third arch) and posterior most part of the tongue is supplied by **vagus nerve** (contribution from superior laryngeal branch of vagus nerve) which is the posttrematic nerve of the fourth arch).

After the tongue is developed, a sulcus, called gingivolingual sulcus appears on either side of the tongue and separates the tongue from the floor of the oral cavity

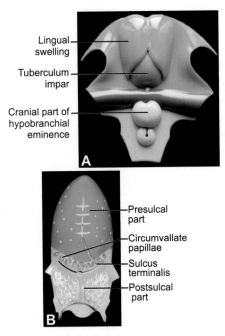

Figs 4.12A and B: Development of tongue

anteriorly leading to freely moving tip and the body of the tongue. The fixed posterior part of the tongue remains as the root of the tongue (Figs 4.12A and B).

98

DEVELOPMENT OF RESPIRATORY SYSTEM

Consists of:

Upper respiratory tract

Lower respiratory tract

- The respiratory system develops after the fourth week of gestation (Figs 4.13A to C).

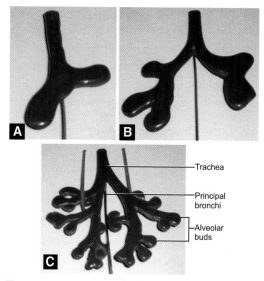

Figs 4.13A to C: Development of respiratory system

Development of Upper Respiratory Tract

Consists of:

Nose, nasopharynx

Begins with the formation of:

- Nasal placode
- Nasal pit
- Nasal sac—Separated from the primitive oral cavity (stomodeum) by oronasal membrane which consists of ectoderm of stomodeum and ectoderm of nasal sac.
- Both nasal sacs are separated by developing nasal septum from posterior part of frontonasal process.

Bucconasal membrane lies behind the primitive palate (premaxilla).

- The oro/bucconasal membrane ruptures behind the palate and nasal cavity communicates with oral cavity behind the secondary palate.
- This communication is the posterior nasal aperture.
- Roof of the nasal cavity is from ectoderm of nasal placode—olfactory epithelium.
- Floor—Definitive palate.

- Lateral wall—Lateral nasal process. From this wall, conchae develop to increase the surface area for conditioning the air in the nasal cavity.
- Medial wall—Frontonasal process—Nasal septum.
- The pharynx develops from cranial part of the foregut.
- The part behind the nasal cavity is called nasopharynx.

Development of Lower Respiratory Tract

- Lower respiratory organs include larynx, trachea, bronchus start developing from 4th week of development.
- Respiratory primordium develops by 28th day as a median out growth from caudal end of ventral wall of primordial pharynx called laryngotracheal groove.
- Tracheobronchial tree develops caudal to 4th pharyngeal pouch.
- Splanchnic mesoderm surrounding the foregut forms connective tissue, cartilage and smooth muscles of respiratory tract.
- By the end of 4th week, laryngeal groove evaginates to form respiratory diverticulum (lung bud).
- Tracheoesophageal folds appear between tracheo-bronchial diverticulum and rest of the pharynx.

- The folds are longitudinal and approach each other to form tracheoesophageal septum which forms partition dividing ventral part as trachea and dorsal part as esophagus.

Development of Larynx

- Epithelium is derived from the endoderm of cranial end of laryngotracheal tube.
- Laryngeal cartilages are derived from mesenchyme of neural crest incorporated in fourth and sixth pharyngeal arches.
- Arytenoid swellings grow towards tongue.
- Primordial glottis is T-shaped laryngeal inlet.
- Epithelium temporarily occludes laryngeal lumen.
- Recanalization starts by 10th week.
- Laryngeal ventricles form during this recanalization.
- Recesses are bounded by folds of mucous membrane called vocal folds and vestibular folds.
- Epiglottis develops from caudal part of hypobranchial eminence.
- Laryngeal muscles develop from 4th and 6th pair of pharyngeal arches which are supplied by vagus nerve
- Growth increases during 1st three years of life.

Development of Trachea

- Endoderm gives rise to epithelium and glands of trachea.
- Splanchnic mesoderm gives rise to cartilage, connective tissue, and muscles of trachea.

Development of Bronchi and Lungs

- Tracheal buds form two out pouchings which are called primary bronchial buds.
- These grow laterally into the pericardioperitoneal canals.
- Segmental bronchi are formed by 7th week.
- Each segmental bronchus and surrounding mass of mesoderm forms primordium of bronchopulmonary segments.
- By 24 weeks about 17 orders of branches have formed and respiratory bronchioles develop.
- Splanchnic mesoderm gives rise to cartilage, smooth muscles, connective tissue and capillaries of the trachea.
- Visceral pleura is derived from splanchnic mesoderm.
- Parietal pleura is derived from somatic mesoderm.
- Pleura develops from the walls of pericardioperitoneal canals.

Maturation of Lungs

Divided into four periods:
- Pseudoglandular period
- Canalicular period
- Terminal saccular period
- Alveolar period.

Pseudoglandular Period

- Lung resembles exocrine gland.
- By 16 weeks all major elements of lung are formed except alveoli.
- Fetus is unable to survive at this stage.

Canalicular Period

- It is between 16-26 weeks.
- Lumen of bronchi and terminal bronchioles are larger
- By 24 weeks, each terminal bronchiole has given 2 or more respiratory bronchioles.
- Alveolar ducts are formed from division of respiratory bronchioles.
- At the end of canalicular period thin-walled terminal saccules are formed.
- Lung tissue is well-vascularized.

Terminal Saccular Period

- Many alveoli develop.
- Epithelium becomes very thin.
- Capillaries bulge into developing alveoli.
- Intimate contact between epithelial and endothelial cells in the form of blood alveoli barrier appear.

Terminal Alveolar Period

- Type 1 alveolar cells are squamous epithelial cells.
- Type 2 are rounded secretory epithelial cells which secrete pulmonary surfactant, a complex mixture of phospholipids.
- Surfactant reduces surface-tension forces and facilitates expansion of terminal saccules by preventing their collapse.
- Fetus of less than 24-26 weeks has surfactant deficiency.

Alveolar Period

- Type 1 alveolar cells become thin.
- Number of capillaries increase.
- Characteristic mature alveoli do not form until after birth. About 95% of alveoli develop postnatally.

DEVELOPMENT OF HEART

- First major system to function.
- Primordial heart and vascular system start in the middle of 3rd week of development.
- Heart functioning starts by early 4th week.
- Cardiovascular system develops from splanchnic mesoderm.

Early Development (Figs 4.14A to D)

- Endothelial strands named angioblastic cords appear in cardiogenic mesoderm by 3rd week.
- Cords canalize and heart tubes are formed.
- Heart starts beating by 22nd-23rd day.
- Blood flow starts by 4th week.
- Primordial heart tube starts developing by 18 days (Figs 4.14A to D).
- Both endothelial lined heart tubes approch and fuse to form single heart tube as a result of lateral folding of embryo.
- Myocardium develops from splanchnic mesoderm called cardiac jelly which is surrounding the pericardial celom.

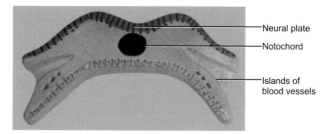

Fig 4.14A

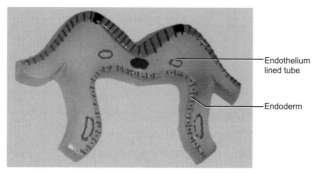

Fig. 4.14B

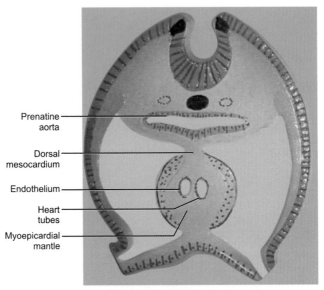

Fig. 4.14C

- Endothelial lining forms endocardium.
- Mesothelial cells form visceral pericardium.
- With the head fold, heart and pericardial cavity come to lie ventral to foregut and caudal to oropharyngeal membrane.

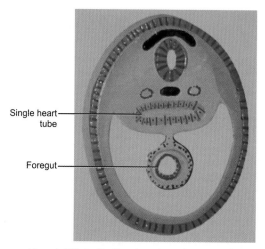

Single heart tube

Foregut

Figs 4.14A to D: Development of heart tube

Gradually heart tube exhibits alternate dilatation and constrictions craniocaudally. These are:

- Truncus arteriosus.
- Bulbus cordis.
- Primitive ventricle.
- Primitive atrium.
- Sinus venosus (Fig 4.15).

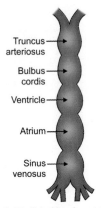

Fig. 4.15: Primitive heart tube

Between dilatations there are endocardial thickenings at constrictions; between sinus venosus and primitive atrium they are called sinoatrial valves; between atrium and ventricle they are termed atrioventricular cushions; between ventricle and bulbus cordis there are proximal bulbar ridges and between bulbus cordis and truncus arteriosus, are distal bulbar ridges.

- Truncus arteriosus is continuous cranially with aortic sac.
- Sinus venosus receives umbilical, vitelline and common cardinal veins from chorion, yolk sac and embryo

110

respectively. These veins open into corresponding horns of sinus venosus.

- Arterial and venous ends of heart tubes are fixed.
- Bulbus cordis and ventricle grow faster and form U-shaped tube called bulboventricular loop.
- Heart elongates and bends, gradually invaginates into pericardial cavity.
- Dorsal mesocardium suspends the heart from the anterior wall of the foregut. Gradually the central part of dorsal mesocardium disappears forming transverse sinus of pericardium (Figs 4.16A to C).

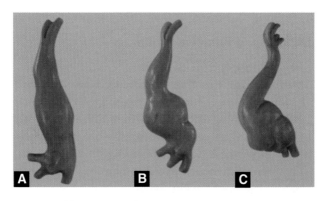

Figs 4.16A to C: Formation of cardiac loop

Partitioning of Primordial Heart

- Partitioning of atrioventricular canal, primitive atrium and primitive ventricle begins around mid 4th week and is completed by 5th week.
- Endocardial cushions appear from dorsal and ventral wall of AV canal and are formed by cardiac jelly.
- Endocardial cushions separate primitive atrium and primitive ventricle, and function as AV valves.
- AV cushions approach each other and fuse to form intermediate AV cushion.
- The formation of intermediate AV cushion results in right and left AV canals.

Partitioning of Primitive Atrium

- Begins at the end of 4th week.
- Septum primum, a thin membrane, descends from the roof of primary atrium to intermediate AV cushion.
- Before this fusion there is a communication between right and left halves of the primitive atrium. It is called foramen primum (Figs 4.17A to D).
- Foramen primum acts as a shunt for oxygenated blood to pass from right to left atrium.

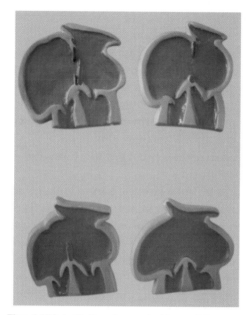

Figs 4.17A to D: Development of interatrial septum

- Septum primum fuses with intermediate endocardial cushion to form primordial interatrial septum.
- Before foramen primum disappears, perforations appear in the central part of septum primum.

- Perforations coalesce to form foramen secundum. foramen primum gradually obliterates.
- A crescent-shaped septum secundum grows from the roof of the primitive atrium to the right of septum primum partially closing the foramen primum and converting it into foramen ovale.
- Septum secundum forms incomplete partition between right and left halves of atrium.
- Lower part of the septum primum forms flap-like valve of foramen ovale.
- Before birth, foramen ovale transmits oxygenated blood from right to left atria. Left to right flow is prevented by septum primum closing on septum secundum.
- After birth foramen ovale closes due to increased pressure in the left atrium after the establishment of pulmonary circulation. Valve of foramen ovale fuses with septum primum thus obliterating the foramen ovale.
- Oval depression in the lower part of interatrial septum of right atrium is a remnant of septum primum and is called fossa ovalis and its margin called limbus fossa ovalis is the remnant of septum secundum fused with septum primum.

Partitioning of Primordial Ventricle

- It is first indicated by a median muscular ridge in the floor of the ventricle near the apex.
- Crescentic fold with concave free edge is formed which forms muscular part of interventricular septum.
- Increase in height of the muscular part of the septum results from dilatation of ventricles on each side of interventricular septum.
- Until 7th week, septum is, crescent-shaped between free edge of interventricular septum and intermediate endocardial cushion.
- Communication between right and left ventricle is through interventricular foramen which closes by the end of 7th week as bulbar ridges fuse with endocardial cushion (Figs 4.18A to C).
- Formation of the membranous part of interventricular septum followed by closure of interventricular foramen, results from fusion of tissues from three sources.
  - The right bulbar ridge.
  - The left bulbar ridge.
  - The intermediate endocardial cushion.
- After closure of interventricular foramen and formation of membranous part of interventricular septum,

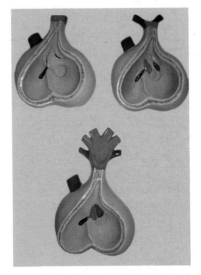

Figs 4.18A to C: Development of interventricular septum

pulmonary trunk communicates with right ventricle, ascending aorta communicates with left ventricle.
- Cavitations of ventricular valves in the form of sponge work of muscular bundle develop into trabeculae carneae.
- Some bundles become the papillary muscles and tendinous cords (chordae tendineae) extending between papillary muscles and AV valves.

116

Aortic Arches and Derivatives (Figs 4.19 to 4.22)

Pharyngeal arches develop during 4th week.

Primitive aortae develop on either side of developing vertebral column. They arch forwards to communicate with primitive heart. The arching takes place in the region of Ist pharyngeal arch.

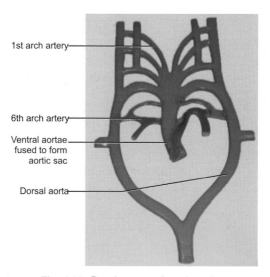

1st arch artery

6th arch artery

Ventral aortae fused to form aortic sac

Dorsal aorta

Fig. 4.19: Development of aortic arches

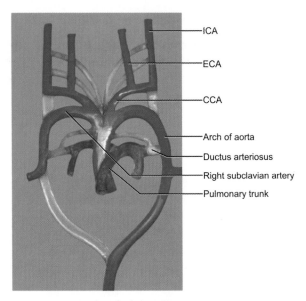

Fig. 4.20: Derivatives of aortic arches

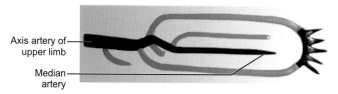

Axis artery of
upper limb

Median
artery

Fig. 4.21: Axis artery of upper limb

Interior

Gluted artery

Fig. 4.22: Axis artery of lower limb

The parts of the primitive aortae on the sides of the vertebral column are called **dorsal aortae**.

The primitive aortae which are extending anteriorly are called **ventral aortae**.

The ventral and dorsal aortae on each side are connected by an arch.

The arched part in the region of Ist pharyngeal arch is the **first aortic arch** or Ist pharyngeal arch artery.

As the caudal pharyngeal arches appear there will be anastomosis between dorsal and ventral aortae of that side lying on the corresponding sides of pharyngeal arches.

Six pairs of aortic arches are formed in which first-two disappear when 6th artery appears.

During 8th week these are transformed into the final fetal arterial arrangement.

Dorsal parts of first aortic arches form **maxillary arteries**.

Derivatives of second pair of aortic arch arteries in its dorsal part persist as stems of **stapedial arteries**.

Derivatives of 3rd pair of aortic arches in the proximal parts form **common carotid arteries** and distal parts join with dorsal aortae to form **internal carotid arteries**.

External carotid arteries appear as buds from 3rd arch arteries.

- Derivatives of 4th aortic arches, on **left side** forms part of **arch of aorta**. On **right side** becomes the proximal part of the **right subclavian artery**.
- Distal part of subclavian artery forms from right dorsal aorta and right **7th cervical intersegmental artery**.
- Left subclavian artery develops from left 7th cervical intersegmental artery.

- Because of differential growth subclavian artery comes to lie close to left common carotid artery.
- Fate of **5th pair** of aortic arch artery: in 50% of embryos 5th pair is **rudimentary** and soon **degenerates** and in other 50% they **do not develop**.
- 6th pair of aortic arches: on left side, proximal part persists as **proximal part** of **left pulmonary artery and distal part** of the arch artery passes from left pulmonary artery to dorsal aorta to form **ductus arteriosus**.
- Right 6th arch artery: proximal part of it persists as **proximal part** of **right pulmonary artery** and **distal part degenerates**.
- Transformation of 6th pair of aortic arches determines the course **of recurrent laryngeal nerves**.
- On the right side as the distal part of right 6th aortic arch degenerates; **right recurrent laryngeal nerve** hooks around **right subclavian artery**, a derivative of 4th aortic arch.
- On the left, **left recurrent laryngeal nerve** hooks around the ductus arteriosus, formed by distal part of 6th aortic arch.
- When arterial shunt involutes after birth the nerve hooks around the **ligamentum arteriosum and the arch of the aorta**.

FETAL CIRCULATION (FIG. 4.23)

- Oxygen and nutrient rich blood from placenta through left umbilical vein enters ductus venosus, which connects left umbilical vein to inferior vena cava.
- Most of the blood from left umbilical vein and vitelline veins bypasses the liver and the remaining portion of blood flows into the sinusoids of the liver. It then enters inferior vena cava through hepatic veins and finally enters right atrium. Through foramen ovale blood enters left atrium and right ventricle.

  From left atrium most of the blood goes to left ventricle from where it reaches aorta; remaining blood goes to lungs.

  The blood from the right ventricle goes to pulmonary trunk; from there blood enters into ductus arteriosus.

- Through ductus arteriosus blood enters descending aorta. From here blood is returned to placenta for reoxygenation via right and left umbilical arteries which are the branches of internal iliac arteries.

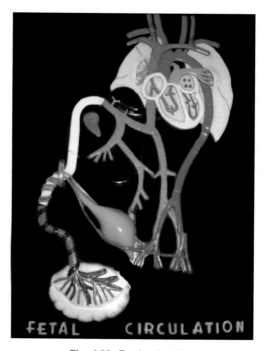

Fig. 4.23: Fetal articulation

Adult derivatives of fetal vascular structures:

- Left umbilical vein—Ligamentum teres of liver.
- Ductus venosus—Ligamentum venosum.
- Umbilical arteries—Proximally become superior vesical arteries and distally medial umbilical ligaments.
- Ductus arteriosus—Ligamentum arteriosum.
- Intra-abdominal part of allantois—Urachus (median umbilical ligament).

Portal Vein

There are three pairs of veins opening into the developing heart (sinus venosus). These are:

1. **Common cardinal veins (ducts of Cuvier)** formed by the fusion of anterior cardinal and posterior cardinal veins which drain the deoxygenated blood from body wall.
2. **Vitelline veins**—Drain the deoxygenated blood from gut and yolk sac.
3. **Umbilical veins**—Drain oxygenated blood from placenta.

Vitelline Veins (Fig. 4.24)

- Vitelline veins start in the yolk sac.
- By 12 weeks the left vitelline vein has regressed and the blood from the left side of the abdominal viscera drains to the right.

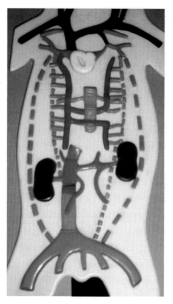

Fig. 4.24: Development of inferior vena cava

125

- The enlarged right vitelline vein between septum transversum and sinus venosus is called right hepatocardiac channel.
- Sinusoids in the liver derived from the vitelline veins and umbilical veins form a final channel which drains into the inferior vena cava.
- Right umbilical vein becomes obliterated during the 2nd month.
- Left umbilical vein anastomoses with the ductus venosus. It is a shunt between left umbilical vein and right hepatocardiac channel (which contributes in the formation of inferior vena cava).

Additionally, the development of the liver from the mesenchymal cells of the septum transversum imposes vascular changes in the process of forming the hepatic sinusoids.

Below the level of septum transversum there will be **anastomosis between right and left vitelline veins** at three levels around the developing duodenum. These anastomoses are—**Cranial ventral, dorsal and caudal ventral**.

The persisting parts of the vitelline veins and the anastomoses give rise to portal vein. They are:

Ventral cranial anastomosis—forms left branch of portal vein.

Right vitelline vein superior to ventral cranial anastomosis—forms right branch of portal vein.

Right vitelline vein between ventral cranial and dorsal anastomosis—forms supraduodenal part of portal vein.

Dorsal anastomosis forms the retroduodenal part of the portal vein.

Cranial part of left vitelline vein between dorsal anastomosis and caudal ventral anastomosis—forms infraduodenal part of portal vein.

Superior mesenteric vein and splenic vein which drain separately into infraduodenal part left vitelline vein get absorbed into the developing portal vein.

Inferior Vena Cava

Components (Fig. 4.24)

- **Hepatic segment**—From right vitelline vein (right hepatocardiac channel).
- **Subhepatic segment**—From anastomosis of right hepatocardiac channel and right subcardinal vein.
- **Prerenal segment**—From right subcardinal vein.
- **Renal segment**—From subcardinal and supracardinal anastomosis.
- **Postrenal segment**—From right supracardinal vein.

- **Postcardinal**—Supracardinal anastomosis.
- **Iliac segment**—From right posterior cardinal vein.

THE GUT DEVELOPMENT

The gut tube develops from endoderm, visceral mesoderm and celomic epithelium including part of yolk sac (Figs 4.25A and B).

- The boundaries between the three parts of the developing gut are indicated by the margins of distribution of the primary arteries of the three parts.
- Extensive elongation results in folding and rotation of the developing gut tube.
- The tube develops through a stage in which the lumen is obliterated and subsequently recanalized.
- Where layers of peritoneum become applied to one another, due to folding, the deep layers are reabsorbed by a process called zygosis.
- Abnormalities of the system are related to the major events of elongation, recanalization, rotation and septation.

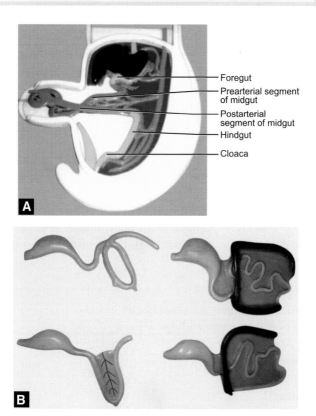

Figs 4.25A and B: Development of gut

FOREGUT

- The foregut develops from the cranial part of the gut. It includes the pharynx, esophagus, stomach, duodenum, associated salivary glands, liver and pancreas.
- It receives its innervation from the glossopharyngeal and vagus nerves and from the cranial and thoracic sympathetic chain.
- Blood supply is by branches of the external carotid artery, thyrocervical trunk, thoracic aorta and celiac trunk.

Venous drainage of the intraabdominal foregut is to the portal venous system, with an anastomosis around the lower esophagus between portal and systemic venous systems.

Derivatives of Foregut

Oral cavity tongue, tonsils, salivary glands.
Primordial pharynx and derivatives.
Upper respiratory system.
Lower respiratory system.
Esophagus.
Stomach.
Duodenum proximal to opening of bile duct.

Liver and biliary apparatus
Lower part of head and uncinate process of pancreas.

Development of Esophagus

The foregut tube is divided rostrally by the tracheo-esophageal septum.

The anterior division develops into the laryngotracheal tube.

The posterior division develops into the esophagus.

The epithelium of the esophagus develops from endoderm.

Proliferation of the epithelium commonly obliterates the lumen (squamous cell metaplasia) but further growth results in recanalization by the end of the embryonic period.

The muscle coat of the esophagus is derived from intraembryonic visceral mesoderm.

Development of Stomach

- The caudal part of the foregut develops into the stomach.
- Initially the primordium of the stomach lies in the midline.
- There will be fusiform dilatation.
- Differential growth of the fusiform dilatation takes place, being faster along the dorsal border than the ventral.

- As this process is occurring, the tube rotates for about 90 degrees – anterior border to the right side and posterior border to the left side.
- The end result of this differential growth is that the cranial end of the stomach lies to the left of the midline, the body lies almost at right angles to the midline, and the pylorus lies to the right of the midline.
- The fast growing part becomes greater curvature slow growing part appears drawn in and forms lesser curvature.

Development of Duodenum

Duodenum develops from caudal most part of foregut and cranial part of prearterial segment of midgut. The changes that occur are:

a. Rotation of gut to 90°.
b. Differential growth—Right wall grows faster and forms 'C' shaped curve.
c. Zygosis—Fusion and absorption of peritoneum on posterior part, pushing it retroperitoneally.

The function between fore- and midgut is opening of hepatopancreatic duct.

Development of Midgut

The midgut forms a U-shaped loop that herniates into the umbilical cord during the 6th week of gestation (physiological umbilical herniation).

While in the umbilical cord, the midgut loop rotates to 90 degrees.

During the 10th week of gestation, the midgut loop returns to the abdomen, rotating to additional 180 degrees.

The cranial limb (prearterial segment) of the midgut elongates rapidly during development and forms the jejunum and cranial portion of the ileum.

The caudal limb (postarterial segment) forms the cecum (Figs 4.26A and B), appendix, caudal portion of the ileum, ascending colon, and proximal two-thirds of the transverse colon.

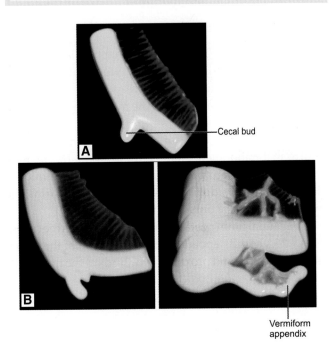

Figs 4.26A and B: Development of cecum

The caudal limb is easily recognized during development because of the presence of the cecal diverticulum.

The midgut loop rotates to 270° counter clockwise around the superior mesenteric artery as it retracts into the abdominal cavity during the tenth week of development.

Midgut Derivatives

Distal duodenum
Jejunum and ileum
Vermiform appendix
Cecum
Ascending colon
Right two-thirds of transverse colon.

The midgut communicates with the yolk sac via the vitellointestinal duct.

Development of Liver

It develops as hepatic diverticulum from the caudal most part of the foregut. It arises from the anterior wall of the endoderm of the foregut in the ventral mesogastrium, towards the septum transversum.

The hepatic diverticulum gives rise to hepatocytes, bile canaliculi and hepatic ducts.

Septum transversum gives rise to connective tissue of the liver.

Sinusoids of the liver are formed by the branching of vitelline and umbilical veins in septum transversum.

Development of Gallbladder

It develops from hepatic diverticulum as pars cystica, which becomes gallbladder and cystic duct.

Development of Pancreas

Develops from endoderm as ventral and dorsal pancreatic buds arising from anterior wall of foregut and posterior wall of midgut at the junction between fore and midguts.

Due to 90 degrees of rotation of foregut, differential growth of duodenal wall and absorption of peritoneum (zygosis) on the dorsal part of the duodenum, the pancreas moves to the left side.

Exocrine Part

Development of duct system:
Main pancreatic duct develops from:

- The duct of the ventral bud.
- Communication between ducts of ventral and dorsal buds.
- The distal part of the duct of the dorsal bud.

Accessory pancreatic duct develops from:
- The proximal part of the duct of the dorsal bud.

The end pieces of the terminal ducts expand to form **acini**.

Endocrine Part of Pancreas

Before canalization of the ducts and formation of acini of exocrine part, clusters of endodermal cells get detached and form islets of Langerhans.

Development of Hindgut

Cloaca (Figs 4.27 and 4.28)

- The caudal dilated part of the hindgut is known as the cloaca. At the end of the cloaca, there is a cloacal membrane which separates the endoderm from the surface ectoderm.

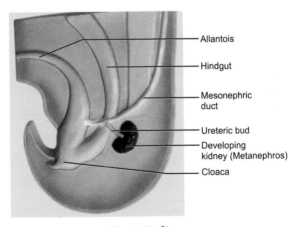

Fig. 4.27: Cloaca

Fig. 4.28: Derivatives of cloaca

Partitioning of Cloaca

The cloaca is the endodermally lined cavity at the end of the gut tube.

It has a diverticulum into the body stalk called the **allantois**.

- The **cloacal membrane** separates the cloaca from the proctodeum (**anal pit**).
- During development a sheet of mesoderm (**urorectal septum**) develops to divide the cloaca into a ventral (**primitive urogenital sinus**) and a dorsal portion (**anorectal canal**).

- By seventh week urorectal septum reaches the cloacal membrane, dividing it into ventral **urogenital membrane** and dorsal **anal membrane**.
- The epithelium of the superior two-thirds of the anal canal is derived from the endodermal hindgut.
- The inferior one-third develops from the proctodeal ectoderm.
- The junction of these two epithelia is indicated by the **pectinate line**, which also indicates the former site of the **anal membrane** that normally ruptures during the **eighth week** of development.

Hindgut Derivatives

Left one-third of transverse colon.

Descending colon, sigmoid, and rectum.

Proximal anal canal (superior to the pectinate line).

The caudal part of the hindgut, known as the cloaca, is divided by the urorectal septum into the urogenital sinus, which gives rise to urogenital system and the primitive rectum, that forms the rectum and proximal anal canal.

UROGENITAL SINUS

Formation and Fate

It is the anterior part of the cloaca.

The mesonephric ducts open into the posterolateral parts of the primitive urogenital sinus dividing the primitive urogenital sinus into **upper vesicourethral canal** and **lower definitive urogenital sinus**. The definitive urogenital sinus is further divide into cranial pelvic part and caudal phallic part.

The vesicourethral canal gives rise to:
- Urinary bladder except trigone in both sexes. Trigone develops from mesonephric ducts.
- Anterior wall and upper part of posterior wall of the prostatic urethra in male.
- Entire female urethra.

Pelvic part of urogenital sinus gives rise to
- Lower part of posterior wall of prostatic urethra and membranous urethra in male, and bulbourethral glands.

Phallic part
- Most of the spongy urethra in male
- Vestibule in female, greater vestibular glands.

URINARY SYSTEM

From intermediate mesoderm—visible at 10 somite stage.

In lower vertebrates—serial segmental diverticuli appear called nephrotomes.

Each nephrotome has a cavity called nephrocele which communicates with peritoneal funnel called nephrostome.

Dorsal wall of nephrotome evaginates as a nephric tubule_which bends caudally at the dorsal tips and fuse with the lower ones to form a longitudinal primary-excretory duct.

This duct curves ventrally and opens into cloaca.

Glomeruli arise from either ventral wall of nephrocele or roof of celom adjacent to the peritoneal funnels or from both.

Renal excretory system is regarded as having 3 organs—pronephros, mesonephros, and metanephros in higher vertebrates.

Pronephros

Present as clusters of cells in cranial part.

Caudally-similar groups become vesicular.

Peritoneal funnels are rudimentary and glomeruli are absent.

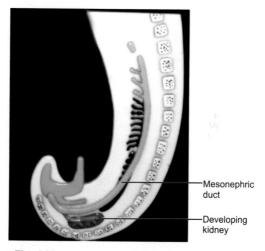

Mesonephric duct

Developing kidney

Fig. 4.29: Formation and fate of nephrogenic cord

Primary excretory duct: A solid rod of cells undergo canalization in the dorsal part of nephrogenic cord (Fig. 4.29). Cranial end is at 9th somite level and caudal end merges in undifferentiated mesenchyme of cord. The duct later elongates and its caudal end becomes detached from the nephrogenic cord to lie immediately beneath ectoderm. From this level it grows caudally- independent of nephrogenic mesoderm and then curves ventrally to reach the wall of cloaca.

Mesonephros

Extent: from septum transversum to 3rd lumbar segment:

The intermediate mesoderm now develops into mesonephric tubule which begins to open into excretory duct-now called mesonephric duct (Wolffian duct).

Mesonephric tubules have been estimated to be 70-80 and gradually glomeruli develop at one end where there is dilatation. The other end of tubule opens into mesonephric duct. The numbers of tubules which are present are not more than 30-40 because when the caudal tubules are developing the cranial ones atrophy.

The whole of the mesonephric tubule now becomes elongated and spindle-shaped and projects into celomic cavity lateral to dorsal mesentery in lower thoracic and lumbar region. It is called mesonephric ridge/Wolffian body. Along with the celomic mesothelium it is called the genital ridge, which gives rise to gonads (Figs 4.30 to 4.32).

The mesonephric ducts open into cloaca.

Just before opening into cloaca the mesonephric duct gives rise to ureteric diverticulum.

It grows towards metanephros to form the ureter, pelvis of kidney, major and minor calyces.

Fig. 4.30:
Development
of gonads

Fig. 4.31:
Development of
testis

Fig. 4.32:
Development of
ovary

Metanephros

It is the permanent kidney, appearing in 5th week of intra-uterine life. Excretory tubules appear, lengthen rapidly to form an S-shaped loop and acquire a tuft of capillaries that will form the glomeruli. Around the glomerulus, excretory tubules form the Bowman's capsule and the rest of the nephron (Fig. 4.33B). Nephrons communicate with the terminal branches of ureteric bud derived from mesonephric duct.

Thus kidney develops from (Figs 4.33A and B):

1. Metanephric blastema that gives rise to secretory part of the kidney.
2. Mesonephric duct (ureteric bud) gives rise to collecting part of the kidney – collecting duct onwards up to ureter.

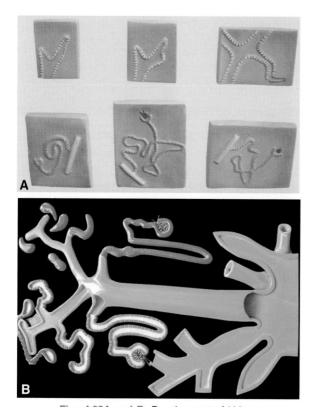

Figs 4.33A and B: Development of kidney

FEMALE REPRODUCTIVE ORGANS—FALLOPIAN TUBES, UTERUS, VAGINA AND OVARY

Fallopian tubes and uterus develop from paramesonephric ducts.

Paramesonephric Ducts (Müllerian Ducts) (Fig. 4.34A)

- Are formed by the invagination of celomic epithelium.
- They lie lateral to the mesonephric ducts in the cranial part of the nephrogenic cord.
- When traced caudally they cross to the medial side of the mesonephric ducts.
- In the midline the right and the left paramesonephric ducts meet and fuse to form uterovaginal canal.
- The caudal ends of this canal come in contact with the dorsal wall of the phallic part of the definitive urogenital sinus.
- This part of the sinus gives rise to vestibule.
- The paramesonephric ducts develop in the female because of a lack of antimüllerian hormone.
- In the male it degenerates, leaving vestigial structures such as the appendix of testis and prostatic utricle.

Development of Uterus

- Fused part of paramesonephric ducts called utero-vaginal canal form the epithelium of the uterus.
- The myometrium is derived from surrounding mesoderm.
- As the thickness of the myometrium increases, unfused horizontal parts of two paramesonephric ducts partially get embedded within its substance to form the fundus.
- The cervix can be recognized as a separate region.
- In the fetus the cervical part is longer than the body of uterus.

Uterine Tubes

- The uterine tubes develop from the unfused parts of paramesonephric ducts.
- The original point of invagination of the paramesonephric ducts in the celomic epithelium remains as the abdominal openings of the tubes. Fimbriae are formed in this situation.

Development of Vagina

The upper part of the vagina develops from the Müllerian ducts.

The lower part of the vagina is formed from the evaginated pelvic part of urogenital sinus (the Müllerian tubercle) which are called sinovaginal bulbs. The sinovaginal bulbs form when the uterovaginal primordium contacts the urogenital sinus.

The lumen of the vagina remains separated from that of the urogenital sinus by a thin tissue plate called the hymen.

Thus vagina develops from:

1. Above hymen – upper four-fifth – the mucosa is derived from sinovaginal bulbs. The musculature develops from the mesoderm of lower fused parts of paramesonephric ducts.
2. Below the hymen – Lower one-fifth – develops from endoderm of urogenital sinus.
3. External vaginal orifice—ectoderm of genital folds after the rupture of urogenital membranes.

Uterus Didelphys (Fig. 4.34B)

It is duplication of uterus resulting from lack of fusion of paramesonephric ducts throughout their normal line of fusion leading to uterus didelphys (Figs 4.34B).

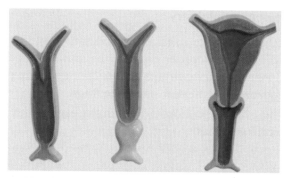

Fig. 4.34A: Development of female reproductive organs

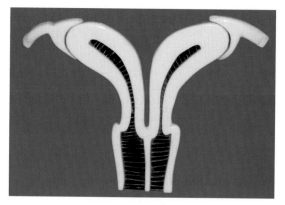

Fig. 4.34B: Didelphis

Development of Ovary

1. Oogonia develop from cells of yolk sac.
2. Follicular cells are derived from the celomic epithelium
3. Connective tissue stroma is from the mesoderm of the gonadal ridge, medial to mesonephros.

The ovaries develop in the lumbar region and later descend into pelvis.

DEVELOPMENT OF TESTIS (FIGS 4.35A AND B)

1. Spermatogonia develop from cells of yolk sac.
2. Basement membrane of the seminiferous tubules and Sertoli cells develop from celomic epithelium.
3. Interstitial cells and connective tissue develop from mesoderm of gonadal ridge, medial to mesonephros.
4. Efferent ductules are derived from the proximal 12 to 15 mesonephric tubules.
5. Mesonephric ducts give rise to canals of epididymis and vas deferens.

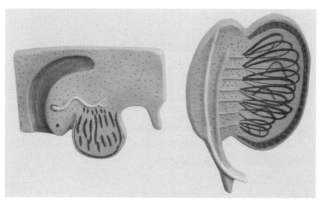

Fig. 4.35A

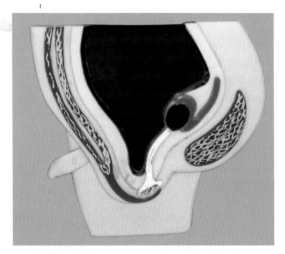

Figs 4.35A and B: Development of testis

6. Cranial end of paramesonephric duct (Müllerian duct) remains as appendix of testis.
7. Cranial end of mesonephric duct (Wolffian duct) remains as appendix of epididymis.

154

Descent of Testis

The testes descend into the scrotal sacs after the development. As the testes are descending, peritoneal covering is pulled along the testis which is called processus vaginalis (Fig. 4.36).

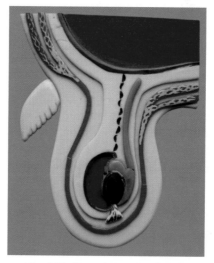

Fig. 4.36: Normal descent of testis

Factors Responsible for the Descent

1. Gubernaculum testis is guiding force of the descent.
2. Increased intraabdominal pressure facilitates the descent.
3. Scrotal temperature is 4 degrees less than the intra-abdominal temperature which favors spermatogenesis.
4. Contraction of the arched fibers of internal oblique muscle.
5. Differential growth of body wall.
6. Testicular hormones secreted by interstitial cells of fetal testes.

   Age and position of testis during its descent:
   4th month of IUL—Iliac fossa.
   7th month of IUL—Deep inguinal ring.
   7th to 8th month of IUL – Traverses inguinal canal.
   9th month of IUL or shortly after birth—Reaches the scrotum.

Anomalies Associated with Descent of Testis (Figs 4.37 to 4.41)

a. **Complete congenital hernia with hydrocele**: Occurs due to complete nonobliteration of processus vaginalis. Coils of intestine descend into the scrotal sac.

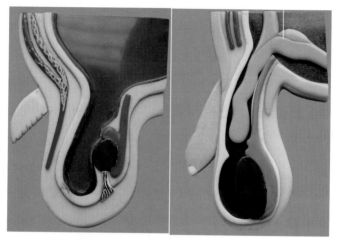

Fig. 4.37: Congenital hydrocele   Fig. 4.38: Complete congenital hernia

b. **Congenital incomplete hernia**: Occurs due to partial obliteration of processus vaginalis in which the coils of intestine do not descend into the scrotal sac.

c. **Infantile hydrocele**: Nonobliteration of processus vaginalis in the distal part (in the scrotal sac).

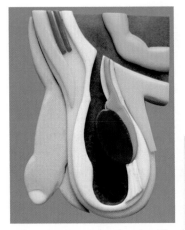

Fig. 4.39: Incomplete hernia with hydrocele

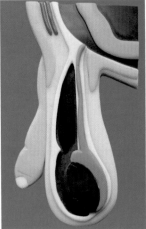

Fig. 4.40: Infantile hydrocele

158

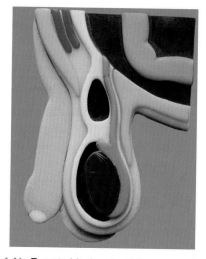

Fig. 4.41: Encysted hydrocele of the spermatic cord

d. **Encysted hydrocele of the spermatic cord**: The processus vaginalis is closed at the deep inguinal ring and at upper part of the testis but it remains patent in the middle.

DEVELOPMENT OF EXTERNAL GENITALIA (FIGS 4.42 TO 4.44)

Male

Development of External Genital Organs

The caudoventral part of urogenital sinus is dilated and is called phallic part of UGS. It is limited by urogenital membrane which is made up of external ectoderm and internal endoderm (Fig. 4.42).

This urogenital membrane is bounded by a elevated margin on either side called genital fold. The genital folds are due to growth and migration of mesoderm at the caudal end of primitive streak. Another elevation called genital tubercle develops in the medial plane, at the cranial part of urogenital membrane. The two genital folds converge at the genital tubercle. Lateral to genital folds, there are another folds called labioscrotal folds or genital swellings.

The genital tubercle elongates and forms phallus. The phallus is invested by ectoderm and consists of mesodermal core. As phallus elongates to form penis, a longitudinal ectodermal groove called primary urethral groove develops

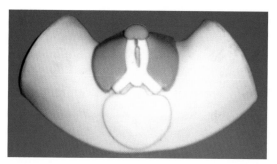

Fig. 4.42: Ext genitalia

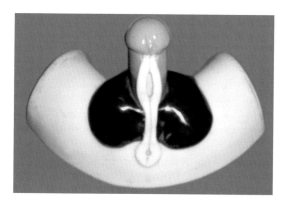

Fig. 4.43: Ext genitalia—male

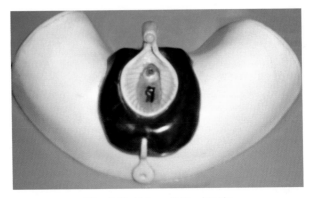

Fig. 4.44: Ext genitalia—female

along the caudal surface of phallus. This ectoderm is continuous with ectodermal cloaca posteriorly. This ectodermal groove extends along the inferior surface of glans penis. A sagittal solid urethral plate of endoderm arises from phallic part of urogenital sinus. It extends forwards within phallus along the primary urethral groove up to the base of glans penis (Fig. 4.43).

The solid urethral plate within phallus undergoes canalization.

The urogenital membrane ruptures and establishes communication between ectodermal canalized urethral plate and endoderm of phallic part of urethra. The genital folds fuse to form spongy part of penis.

The genital (labioscrotal) swellings give rise to scrotum.

Female

The caudoventral part of urogenital sinus is dilated and is called phallic part of UGS. It is limited by urogenital membrane which is made up of external ectoderm and internal endoderm.

This urogenital membrane is bounded by elevated margin on either side called genital fold. The genital folds are due to growth and migration of mesoderm at the caudal end of primitive streak. Another elevation called genital tubercle develops in the medial plane, at the cranial part of urogenital membrane. The two genital folds converge at the genital tubercle. Lateral to genital folds, there are another folds called labioscrotal folds or genital swellings (Figs 4.42 to 4.44).

The genital tubercle in the midline, caudoventrally gives rise to clitoris.

The genital folds become labia minora.

The genital swellings (Labioscrotal swellings) become the labia majora.

Summary of Urogenital System

Embryonic structure	Male	Female
Indifferent gonad	Testis	Ovary
Cortex	Seminiferous tubules	Ovarian follicles
Medulla	Rete testis	Rete ovarii
Gubernaculum	Gubernaculum testis	Ovarian and round ligament of uterus
Mesonephric tubules	Ductuli efferentes	Epoophoron, paroophoron
Mesonephric duct	Appendix of epididymis, duct of epididymis, ductus deferens, ureter, calices and collecting tubules, ejaculatory duct and seminal vesicles	Appendix vesiculosa, duct of epoophoron, duct of Gartner, ureter, pelvis, calices and collecting tubules
Paramesonephric duct	Appendix of testis	Hydatid of Morgagni, uterine tube, uterus
Urogenital sinus	Urinary bladder, urethra, prostatic utricle, prostate gland and bulbourethral glands	Urinary bladder, urethra, vagina, urethral, paraurethral and greater vestibular glands
Sinus tubercle	Seminal colliculus	Hymen
Phallus	Penis	Clitoris
Urogenital folds	Ventral aspect of penis	Labia minora
Labioscrotal swellings	Scrotum	Labia majora

164

DEVELOPMENT OF NERVOUS SYSTEM AND SPECIAL SENSES

Neurulation

Neural plate—Ectodermal cells overlaying the notochord become tall columnar, producing a thickened neural plate (in contrast to surrounding ectoderm that produces epidermis of skin), by the 17-20th day of gestation.

Neural groove—The neural plate is transformed into a neural groove by the 23rd day.

Neural tube—The dorsal margins of the neural groove merge medially, forming a neural tube composed of columnar neuroepithelial cells surrounding a **neural cavity** on the 25th day. The neural tube has two openings—a cranial opening called anterior neuropore and a caudal opening called the posterior neuropore. The closure of the neural tube is dependent upon protein bridges bound together by calcium.

In the process of separating from overlaying ectoderm, some neural plate cells become detached from the tube and collect bilateral to it, forming **neural crest**.

By 25-26th day the anterior neuropore closes and by 27th day the posterior neuropore closes.

Once closure is effected, the neural crest also begins to form.

The crest is the source of neurons for the peripheral nervous system as well as for chromaffin cells in the inner part of the adrenal gland. Chromaffin cells are responsible for synthesizing and secreting two important hormones instrumental in emotional arousal—epinephrine and norepinephrine.

The neural crest is derived from neuroectodermal cells that originate in the dorsal aspect of the neural folds or neural tube; these cells leave the neural tube or folds and differentiate into various cell types including dorsal-root ganglion cells, autonomic ganglion cells, the chromaffin cells of the adrenal medulla, Schwann cells, sensory ganglia cells of cranial nerves, 5, 9, and 10, part of the meninges and integumentary pigment cells.

Derivatives of Neural Tube and Neural Crest

- Neural tube will form brain and spinal cord.

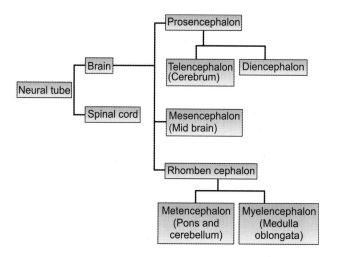

- Neural crest derivatives are: Ganglia of dorsal root, sensory ganglia of V, VII, VIII, IX, X cranial nerves and autonomic ganglia.
- Enamel of tooth.
- Melanocytes.
- Adrenal medulla.
- Parafollicular cells of thyroid.
- Schwann cells.

HISTOGENESIS OF NERVOUS TISSUE

At first, the wall of the neural tube is composed of a single layer of columnar ectodermal cells. Soon, the side-walls become thickened, while the dorsal and ventral parts remain thin, and are named the **roof-** and **floor-plates** (Fig. 4.45).

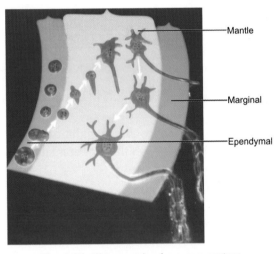

Mantle

Marginal

Ependymal

Fig. 4.45: Histogenesis of nervous system

- A transverse section of the tube at this stage presents an oval outline, while its lumen has the appearance of a slit. The cells which constitute the wall of the tube proliferate rapidly, lose their cell-boundaries and form a syncytium. This syncytium consists at first of dense protoplasm with closely packed nuclei, but later it opens out and forms a looser meshwork with the cellular strands arranged in a radiating manner from the central canal.

Three layers may now be defined—An internal or ependymal, an intermediate or mantle, and an external or marginal.

- The **ependymal layer** is ultimately converted into the ependyma of the central canal; the processes of its cells pass outward toward the periphery of the medulla spinalis.
- The **marginal layer** is devoid of nuclei, and later forms the supporting framework for the white funiculi of the medulla spinalis.
- The **mantle layer** represents the whole of the future gray columns of the medulla spinalis; in it the cells are differentiated into two sets:
  a. **Spongioblasts** or **young neuroglia cells**
  b. **Germinal cells,** which are the parents of the **neuroblasts** or **young nerve cells**.

The triple layered arrangement seen in the spinal cord (ependymal, mantle and marginal layers) continues into the caudal portion of the brain (the medulla oblongata), but becomes more complex cranially. In the brain, the neuronal cell bodies comprising the gray matter become clustered into groups called **nuclei**.

In some parts of the brain, neurons and neuroglia differentiating from the mantle layer of the original neural tube migrate outwards through the white matter of the marginal layer and they form a peripheral, multilayered covering of gray matter. This outer covering of gray matter on the cerebral and cerebellar hemispheres is the cerebral and cerebellar cortex.

By subdivision the germinal cells give rise to the neuroblasts or young nerve cells, which migrate outward from the sides of the central canal into the mantle layer and neural crest, and at the same time become pear-shaped; the tapering part of the cell undergoes still further elongation, and forms the axis cylinder of the cell.

Neurons develop from neuroblasts of the neuroepithelium and migrate into the mantle layer. The neuroblast develops into a bipolar cell having a primitive axon and dendrite. The single dendrite degenerates and is replaced by multiple dendrites forming a multipolar neuroblast. Axons have few branches and their axonal growth cones are directed to their targets by tropic factors. Primitive supporting cells are glioblasts and migrate into the mantle and marginal layers to become astrocytes and oligodendrocytes. Microglia are derived from invading blood monocytes.

- Many neuroblasts migrate (travel) like amoeba, that is, by extending a part of itself, grabbing something to hang onto and then pulling the rest of the cell along. In the neocortex and cerebellum, neuroblasts must travel to their final destinations, locating themselves in the correct cell layer, orient themselves, and initiate dendritic growth to make the appropriate synaptic connections.

- The lateral walls of the medulla spinalis continue to increase in thickness, and the canal widens out near its dorsal extremity, and assumes a somewhat lozenge-shaped appearance. The widest part of the canal serves to subdivide the lateral wall of the neural tube into a **dorsal** or **alar**, and a **ventral** or **basal lamina**.

- In the third and fourth week the development of the spinal cord is seen, and by the end of the fourth week of gestation, the marginal layer nerve fibers appear and begin to accept fibers of ganglion, nerves that are sent into them from the peripheral ganglia. Once connected, they begin to function.

- Fifth week—The cerebral hemispheres differentiate.
- By the end of the sixth week the rudimentary development of the five brain vesicles is complete. The cerebral hemispheres have grown and now cover the diencephalon, the mesencephalon and the cerebellum. As these two hemispheres grow toward each other, they meet in the middle and continue their growth downwards. The membrane that separates them is the falx cerebri—a part of the dural membrane system of the meninges, of which it is the outer layer—the dura mater. The fissure thus created is known as the longitudinal fissure (Figs 4.46 to 4.51).
- By the seventh week of life, the pineal gland, choroid plexus are formed from the roof of the diencephalon and their specialized cells secrete cerebrospinal fluid. Rhombic lip gives rise to cerebellum.

The part of the neural tube which is forming Pons and open part of medulla oblongata consists of laterally placed alar laminae which are separated from the basal laminae by sulcus limitans.

Each basal lamina contains somatic and visceral motor neurons. Each alar lamina contains association neurons

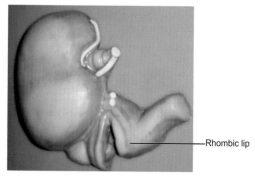

—Rhombic lip

Fig. 4.46: Development of brain—Rhombic lip

which synapse with afferent fibers from sensory neurons of dorsal root gangila.

By 28th day all motor nuclei of cranial nerve are distinguishable. Each basal lamina divides into three columns and each alar lamina divides into four columns. They are as follows from medial to lateral:

BASAL LAMINA

1. Somatic efferent column
2. Special visceral efferent column
3. General visceral efferent column

Columns	Cranial nerves of brain stem	Area of supply
Somatic efferent	3rd, 4th, 6th and 12th	Extraocular muscles and tongue muscles
Special visceral efferent	5th, 7th, 9th, 10th and 11th	Striated muscles of pharyngeal arches
General visceral efferent	Parasympathetic nerves–Edinger Westphal nucleus of 3rd Cr. N, 7th 9th and dorsal nucleus of 10th Cr. nerves	Sphincter pupillae, ciliaris muscle, smooth muscle and glands of cardiovascular system, respiratory system, alimentary canal, pelvic organs and salivary glands.

THE ALAR LAMINA

1. General visceral afferent column
2. Special visceral afferent column
3. General somatic afferent column
4. Special somatic afferent column

Columns	Cranial nerves associated in brain stem	Area of sensation
General visceral afferent	9th and 10th	Sensory receptors from thorax, abdomen and pelvic viscera
Special visceral afferent	Nucleus tractus solitarius, salivatory nucleus, 7th, 9th and 10th	Tongue-taste
General somatic afferent	5th, 7th, 9th and 10th	Skin of head and neck, mucosa of oral nasal cavities and pharynx and laryngeal
Special somatic afferent	8th	Vestibule and cochlea (Special sense of balance and hearing)

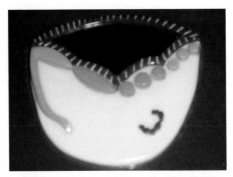

Fig. 4.47: Open part of medulla oblongata (brain stem)

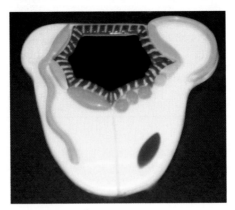

Fig. 4.48: Pons

In medulla oblongata, the inferior olivary nucleus is from alar lamina.

In Pons the alar lamina migrates ventrally to form pontine nuclei.

In mid brain, alar lamina gives rise to collicular nuclei and substantia nigra. Basal lamina gives rise to red nuclei 3rd and 4th cranial nerve nuclei.

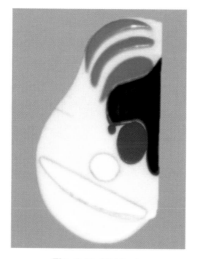

Fig. 4.49: Mid brain

177

In spinal cord, basal lamina gives rise to anterior gray column (motor) and alar lamina gives rise to posterior gray column (sensory).

The lateral part of basal lamina in thoracolumbar and sacral regions give rise to sympathetic and parasympathetic gray column (special visceral efferent column) respectively.

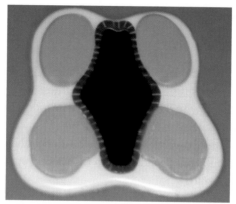

Fig. 4.50: Development of spinal cord

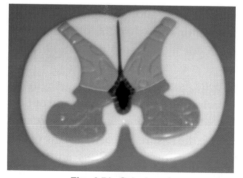

Fig. 4.51: Spinal cord

179

SPECIAL SENSES

Formation of the Eye

- Both eyes are derived from a single field of the neural plate. The single field separates into bilateral fields associated with the diencephalon. The following events produce each eye (Figs 4.52 to 4.56):
  - A lateral diverticulum from the diencephalon forms an *optic vesicle* attached to the diencephalon by an *optic stalk*.
  - A *lens placode* develops in the surface ectoderm where it is contacted by the optic vesicle.

  The lens placode induces the optic vesicle to invaginate and form an **optic cup** while the placode invaginates to form a *lens vesicle* that invades the concavity of the optic cup.
  - An *optic fissure* is formed by invagination of the ventral surface of the optic cup and optic stalk, and hyaloid *artery, a branch of central artery of retina,* invades the fissure to reach the lens vesicle.
- The optic cup forms the *retina* and contributes to formation of the *ciliary body* and *iris*. The outer wall of

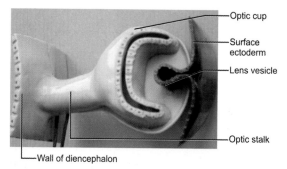

Fig. 4.52: Development of eyeball

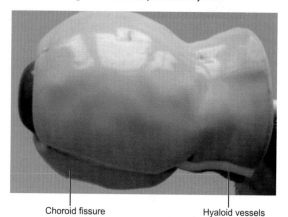

Fig. 4.53: Choroid fissure

181

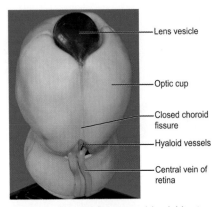

Lens vesicle

Optic cup

Closed choroid fissure

Hyaloid vessels

Central vein of retina

Fig. 4.54: Choroid fissure and hyaloid artery

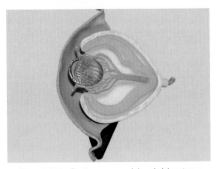

Fig. 4.55: Optic cup and hyaloid artery

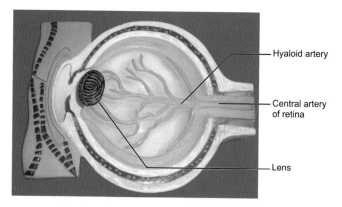

Fig. 4.56: Layers of the fully developed eyeball

the optic cup forms the *outer pigmented layer of the retina*, and the inner wall forms *neural layers of the retina*.

- The optic stalk becomes the optic nerve as it fills with axons traveling from the retina to the brain.
- The lens vesicle develops into the lens, consisting of layers of lens fibers enclosed within an elastic capsule.
- The vitreous compartment develops from the concavity of the optic cup, and the *vitreous body* is formed from

ectomesenchyme that enters the compartment through the optic fissure.

- Ectomesenchyme (from neural crest) surrounding the optic cup condenses to form inner and outer layers, the *future choroid and sclera*, respectively.
- The *ciliary body* is formed by thickening of choroid ectomesenchyme plus two layers of epithelium derived from the underlying optic cup; the ectomesenchyme forms *ciliary muscle and the collagenous zonular fibers* that connect the ciliary body to the lens.
- The *iris* is formed by choroid ectomesenchyme and the superficial edge of the optic cup.
- The outer layer of the cup forms *dilator and constrictor muscles* and the inner layer forms pigmented epithelium.
- The ectomesenchyme of the iris forms a *pupillary membrane* that conveys an anterior blood supply to the developing lens; when the membrane degenerates following development of the lens, a pupil is formed.
- The *cornea* develops from two sources:
  1. The layer of ectomesenchyme that forms sclera is induced by the lens to become inner epithelium and stroma of the cornea.

2. While surface ectoderm forms the outer epithelium of the cornea.

- The *anterior chamber* of the eye develops as a cleft in the ectomesenchyme situated between the cornea and the lens.

- The *eyelids* are formed by upper and lower folds of ectoderm, each fold includes a mesenchyme core; the folds adhere to one another but they ultimately separate either prenatally (ungulates) or approximately two weeks postnatally (carnivores).

- Ectoderm lining the inner surfaces of the folds becomes *conjunctiva, and lacrimal glands* develop by budding of conjunctival ectoderm.

- *Skeletal muscles* that move the eye (extraocular eye muscles) are derived from rostral somitomeres

- (Innervated by cranial nerves III, IV, and VI).

Formation of the Ear (Figs 4.57A and B)

- The ear has three components: External ear, middle ear, and inner ear.
- The inner ear contains:
  1. Sense organs for hearing (cochlea)
  2. Detecting head acceleration (vestibular apparatus), the latter is important in balance.
- The middle ear contains bones (ossicles) that convey vibrations from the tympanic membrane (ear drum) to the inner ear.
- The outer ear.

Inner Ear

- An otic placode develops in surface ectoderm adjacent to the hindbrain in the floor of the first pharyngeal cleft.
- The placode invaginates to form a cup which then closes and separates from the ectoderm, forming an otic vesicle (otocyst).
- Otic capsule, composed of cartilage, surrounds the otocyst.

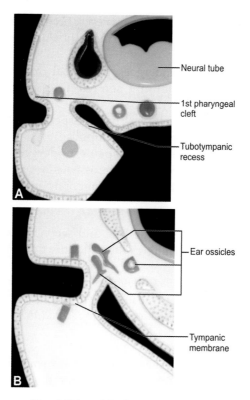

Figs 4.57A and B: Development of ear

- Some cells of the placode and vesicle become neuroblasts and form afferent neurons of the vestibulocochlear nerve (VIII).
- The otic vesicle undergoes differential growth to form the cochlear duct and semicircular ducts of the membranous labyrinth.
- Some cells of the labyrinth become specialized receptor cells found in maculae and ampullae.
- The cartilaginous otic capsule undergoes similar differential growth to form the osseous labyrinth within the future petrous part of the temporal bone.

Middle Ear

- The dorsal part of the first and second pharyngeal pouchs form the lining of the auditory tube, tympanic cavity.
- The *malleus and incus* develop as endochondral bones from ectomesenchyme in the *first branchial arch* and the *stapes* develops similarly from the *second arch*.

Outer Ear

- The tympanic membrane is formed by apposition of endoderm and ectoderm where the first pharyngeal pouch is apposed to the cleft between the first and second branchial arches.

- The external auditory meatus is formed by the groove between the first and second branchial arches. The arches expand in the form of auditory hillocks laterally to form the wall of the canal and the auricle (pinna) of the external ear.

The Taste

Taste Buds

- The taste buds become apparent during the eighth week of gestation, and by the fourteenth week the taste sensation is formed. At birth, infants express positive and aversive facial responses to tastes.
- Taste buds are groups of specialized (chemoreceptive) epithelial cells localized on papillae of the tongue. Afferent innervation is necessary to induce taste bud formation and maintain taste buds. Cranial nerves VII (rostral two-thirds of tongue) and IX (caudal third of tongue) innervate the taste buds of the tongue.

Olfaction (The Sense of Smell)

- Olfaction (smell) involves olfactory mucosa located caudally in the nasal cavity and the vomeronasal organ located rostrally on the floor of the nasal cavity. Olfactory neurons are chemoreceptive; their axons form olfactory nerves (I).

- An olfactory (nasal) placode appears bilaterally as an ectodermal thickening at the rostral end of the future upper jaw; the placode invaginates to form a nasal pit that develops into a nasal cavity as the surrounding tissue grows outward; in the caudal part of the cavity, some epithelial cells differentiate into olfactory neurons;

- The vomeronasal organ develops as an outgrowth of nasal epithelium that forms a blind tube; some epithelial cells of the tube differentiate into chemoreceptive neurons.

Sensory Development—Touch

The somatosensory system begins to develop during gestation. The nervous system, which is the message carrier to the brain for the senses, begins to develop at the third week of gestation. At the ninth week of gestation the sensory nerves have developed and are touching the skin. By the twenty-second week of gestation, the fetus is sensitive to touch and temperature. At birth, the sense of touch can be observed through the infant's reflexes when it comes in contact with different stimuli.

DEVELOPMENT OF ENDOCRINE GLANDS

Development of Pituitary Gland

It develops from two sources. They are (Figs 4.58A and B):
1. Ectodermal roof of stomodeum
2. Neuroectodermal diverticulum from the floor of the third ventricle (hypothalamus).

The ectodermal roof of the stomodeum grows cranially towards the base of the diencephalon as a diverticulum. It is called Rathke's pouch.

The pouch becomes elongated and gets separated from the ectodermal stomodeum to form vesicle. This vesicle approaches the diverticulum from the floor of the diencephalon and surrounds it.

The anterior wall of the vesicle grows more and becomes pars distalis, the tubular prolongation of the vesicle forms pars tuberalis and the posterior wall becomes the pars intermedia.

The down growth from the diencephalon retains its continuity and becomes pars nervosa, pars infundibularis and median eminence which is part of hypothalamus. Cavity of Rathke's vescicle becomes intermediate cleft.

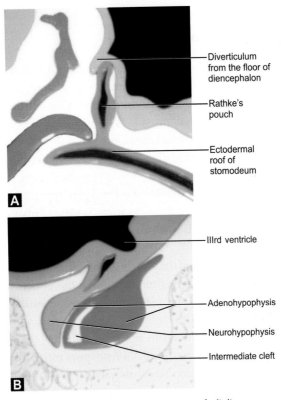

Figs 4.58A and B: Development of pituitary

Development of Suprarenal Glands

The gland consists of:

a. Outer cortex

b. Inner medulla.

Development of suprarenal gland is as follows (Figs 4.59A to C).

The adrenal gland begins to develop in the V week of IUL.

The cells of **cortex** arise from **celomic epithelium** that lies in the angle between the developing gonad and mesentery.

The cells of medulla are derived from **neural crest**.

Development of Cortex

The cells that are formed first are large and acidophilic and surround the medulla, they form fetal cortex. The cortex engulfs the medulla until the medulla is completely encapsulated. The fetal cortex disappears after birth.

The celomic epithelium gives rise to small cells that surround the fetal cortex. This is the definitive cortex.

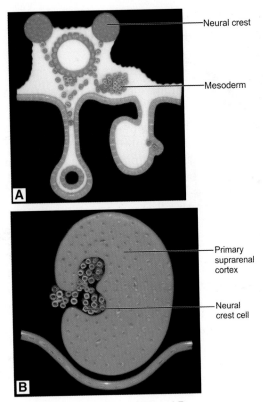

Figs 4.59A and B

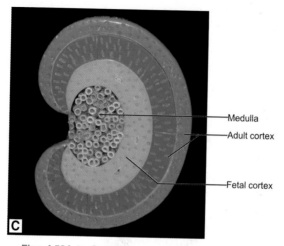

Figs 4.59A to C: Development of suprarenal gland

Development of Medulla

The cells of medulla are derived from neural crest.

These cells migrate to the region of developing cortical cells and come to be surrounded by them.

- Throughout the remaining fetal period, two cortical layers surround the gland, first the zona glomerulosa and then the zona fasciculata.

195

- A final layer, the zona reticularis forms after birth at about 3 years.
- The fetal adrenal glands are relatively large.
- At four months of intrauterine life the glands are larger than the fetal kidney.
- At birth they are relatively large (about 20 times their relative size in the adult).
- After birth, the fetal cortex rapidly undergoes involution (regression).
- It is the inner "fetal zone" of the adrenal cortex that persists during gestation, suggesting that these cells are involved in a specific function during pregnancy.

Development of Thyroid Gland

It develops as a midline endodermal diverticulum from the floor of the developing foregut at the developing tongue region. It is called the midline thyroid diverticulum. It grows up to the level of 4th pharyngeal pouches. The elongated stalk is called the thyroglossal duct and the caudal end of the duct there is thyroid diverticulum. This grows into lobes and the isthmus. The lobes are reinforced by the ventral parts of the fourth pouches called the lateral thyroid diverticula. The parafollicular cells of thyroid gland develop from ultimobranchial bodies which contain the incorporated neural crest cells.

The connective tissue and vessels develop from the visceral mesoderm.

Development of Parathyroid Glands

They are four in number. On either side of the midline. Two are superior and two are inferior.

Superior parathyroids develop from endoderm of the dorsal part of fourth pouches.

And inferior parathyroids are from the endoderm of the dorsal part of third pouches.

The third migrate inferior to fourth pouches along with the migrating thymic lobes which bring them below the level of the fourth pouches.

The connective tissue and the vessels develop from the surrounding visceral mesoderm.

Index